Mathematics of Multilevel Systems

Data, Scaling, Images, Signals, and Fractals

Contemporary Mathematics and Its Applications: Monographs, Expositions and Lecture Notes

Print ISSN: 2591-7668
Online ISSN: 2591-7676

This series aims to inspire new curriculum and integrate current research into texts. Its aims and main scope are to publish:

- Cutting-edge Research Monographs
- Mathematical Plums
- Innovative Textbooks for capstone (special topics) undergraduate and graduate level courses
- Surveys on recent emergence of new topics in pure and applied mathematics
- Advanced undergraduate and graduate level textbooks that may initiate new directions and new courses within mathematics and applied mathematics curriculum
- Books emerging from important conferences and special occasions
- Lecture Notes on advanced topics

Monographs and textbooks on topics of interdisciplinary or cross-disciplinary interest are particularly suitable for the series.

Published

Vol. 8   *Mathematics of Multilevel Systems: Data, Scaling, Images, Signals, and Fractals*
           by Palle E T Jorgensen & Myung-Sin Song

Vol. 7   *Operator Theory and Analysis of Infinite Networks: Theory and Applications*
           by Palle E T Jorgensen & Erin P J Pearse

Vol. 6   *Generalized Radon Transforms and Imaging by Scattered Particles:*
           Broken Rays, Cones, and Stars in Tomography
           by Gaik Ambartsoumian

More information on this series can also be found at
https://www.worldscientific.com/series/cmameln

Contemporary Mathematics and Its Applications
Monographs, Expositions and Lecture Notes

Vol. **8**

Mathematics of Multilevel Systems

Data, Scaling, Images, Signals, and Fractals

Palle E T Jorgensen
The University of Iowa, USA

Myung-Sin Song
Southern Illinois University Edwardsville, USA

World Scientific

EW JERSEY · LONDON · SINGAPORE · BEIJING · SHANGHAI · HONG KONG · TAIPEI · CHENNAI · TOKYO

Published by

World Scientific Publishing Co. Pte. Ltd.

5 Toh Tuck Link, Singapore 596224

USA office: 27 Warren Street, Suite 401-402, Hackensack, NJ 07601

UK office: 57 Shelton Street, Covent Garden, London WC2H 9HE

Library of Congress Cataloging-in-Publication Data

Names: Jørgensen, Palle E. T., 1947–   author. | Song, Myung-Sin, author.
Title: Mathematics of multilevel systems : data, scaling, images, signals, and fractals /
　　Palle E.T. Jorgensen, The University of Iowa, USA,
　　Myung-Sin Song, Southern Illinois University Edwardsville, USA.
Description: New Jersey : World Scientific Publishing Co. Pte. Ltd., [2023] |
　　Series: Contemporary mathematics and its applications. Monographs, expositions and
　　lecture notes, 2591-7668 ; Vol. 8 | Includes bibliographical references and index.
Identifiers: LCCN 2022053824 | ISBN 9789811268977 (hardcover) |
　　ISBN 9789811268991 (ebook for institutions) | ISBN 9789811269011 (ebook for individuals)
Subjects: LCSH: Wavelets (Mathematics) | Hilbert space. | Frames (Combinatorial analysis)
Classification: LCC QC20.7.W38 J67 2023 | DDC 515/.2433--dc23/eng20230324
LC record available at https://lccn.loc.gov/2022053824

British Library Cataloguing-in-Publication Data
A catalogue record for this book is available from the British Library.

For any available supplementary material, please visit
https://www.worldscientific.com/worldscibooks/10.1142/13227#t=suppl

Desk Editors: Aanand Jayaraman/Lai Fun Kwong

Typeset by Stallion Press
Email: enquiries@stallionpress.com

Preface

Recent experience in multiple mathematics departments around the country has suggested that, in addition to classes, an important learning experience for students is a variety of guided projects. And the present book is inspired in part by such senior and master's projects for students. While there are many ways students may be exposed to interactive projects, we stress here the topics, and the mathematical framework for such hands-on areas, as the following interrelated themes: wavelet algorithms, image processing, image compression, dimension reduction, lifting schemes, analysis of signals, the study of similarity up to scale, dynamics, and of notion of self-similarity as it arises in the study of fractals. Indeed, these related topics have proved well-suited for a variety of senior and master's projects, aimed at undergraduate and beginning graduate students.

Historically, wavelet algorithms have proved extremely successful as compared to alternative harmonic analysis tools. They have been picked up by engineers at an impressive rate (especially by electrical, computer and industrial engineers). The adoption by engineers has been truly impressive in both areas of teaching, as well as in such industrial applications as data mining, machine learning, neural networks, and digital images.

Our presentation in this book will cover both the mathematics as well as neighboring applied themes, which are the focus of this volume. We further aim to make the subject accessible to students by covering in equal measure the following three themes: (i) a selection of practical projects and algorithms, (ii) the theory underpinning the subjects, and (iii) also the important interplay between theory and applications. In order not to interrupt the flow of topics inside the book, we have relegated some background-theory topics to Appendices A through D: Hilbert Space (A),

Wavelet Algorithms (B), Cantor Dynamics (C), Markov Chains and Multiresolutions (D).

Thus, the success of wavelet algorithms has been borne out in a rich variety of applications; most notably (stressed here) their use in design of algorithms for processing, transmission, and storing, of digital signals — and images. The latter include those used in storage of fingerprints, in digital form. Fundamental to both the mathematics and the engineering side is the notion of *multiresolution* (i.e., a sequential algorithm for quantization of *Resolution* and *Detail features*). It serves as a basic and versatile tool in dealing with data; in "translating" raw data into scales of multiresolution features. It serves in particular as a powerful tool in the design of digital algorithms, and it will be stressed in our present exposition. We use it as a universal tool, in the processing of *Resolution* and *Detail* data, in each step of the algorithm, but algorithms will be adapted to the hands-on problems. Since the initial pioneering papers and books on wavelet theory (as part of modern harmonic analysis), the areas where wavelet algorithms have found success have expanded greatly, and in multiple and diverse directions.

Here, we shall stress the following (not an exhaustive list): fractals, chaos theory, noise detection, dynamical systems, specifically the study of path space, of measures on solenoids; and adaptation of multiresolution scales to *Julia Sets* (from complex dynamics).

The purpose of the present book is to make these diverse and powerful tools more accessible to students. This will be accomplished with new viewpoints, and with hands-on projects; ideas and projects we have tested with students. One of our aims has been to make the connection between multiresolutions as they arise in both mathematics, and in their incarnations in applications, e.g., in the processing of the algorithm to grayscale numbers in digital images. (In color images, the algorithm further entails a mix of the three basic colors.) We hope that our presentation will be student-friendly. It is inspired by a variety of student projects we have directed over the past decade.

Our book entails a varied choice of diverse interdisciplinary themes. While the topics can be found in various parts of the pure and applied literature, we felt that there is a need for an accessible presentation which cuts across fields. So, in particular, we have aimed at "translating" between lingo which tends to be popular in different areas; thus fleshing out "translations" from lingo used in one area to that of other areas.

As a result, our target audience is diverse as well, both with regards to level and to choice of topics. On the traditional mathematics side, we aim to reach users, at multiple levels, and with diverse interests, students and others who want to learn about exciting connections to engineering problems involving wavelet mathematics.

On the applications side, our target audience includes, among others, engineers (both students and others) who want to catch up on the mathematics which is an integral part of a number of new engineering developments, but which also relies on core mathematics, and on key tools from wavelet algorithms.

A Note on Infinite Dimensions: Since our target audience is diverse, we have postponed a detailed/systematic discussion of issues involving infinite dimensions and Hilbert space until Chapters 3 (Wavelets), 4 (Transform Theory), and 5 (Entropy Encoding and Probability). For the problems addressed there, the case of infinite dimension will be more natural and well-motivated. Hence, we have chosen to postpone a systematic presentation of Hilbert space theory.

As for our present preliminary discussion of image resolutions, finite dimensions will suffice. And the choice of treating linear algebra tools first will further serve as an introduction to the more advanced questions from later chapters. There we treat operators in Hilbert space, and include rigorous definitions. But for the present purpose, suffice it to call attention to the following intuition behind matrix operations used on image resolutions (see, e.g., Chapter 2), and involving only finite-dimensional data sets. Specifically, as the digital image under consideration is successively refined, via the use of iteration of data points for scaling and detail, then of course the finite-dimensional data sets involved will increase, with each iteration of the image resolution algorithm. The resulting issues of limits will be addressed in later chapters. They deal with precise notions of "limits." In rough outline, finite-dimensional linear algebra, in the limit, transitions into infinite-dimensional analysis, and entails the use of operators in Hilbert space.

As mentioned above, our target audience is diverse, both with regards to subject areas and applications as well as levels. Indeed, we are hoping to reach both beginners, as well as experienced students. In fact, this diversity of our target audience is by design. And so, for this reason, when new notions are introduced inside the book, they will be carefully explained from first principles. In addition, we have included four special **student-tools.** They are as follows: (i) **Glossary** (Section 3.2), (ii) **List of Names and Discoveries** (Section 4.5), and (iii) **Multiple Tables and Figures** throughout the book, each one serving to illustrate main ideas, algorithms,

Mathematics of Multilevel Systems

and results. In addition, we have endeavored (as far as possible) to make the chapters independent; and we expect that most readers will be able, without too much difficulty, to pick up the thread in any one of the chapters inside the book.

We hope these features will help take the fear out of otherwise intimidating math terminology, used inside the chapters of the book; e.g., our use of such key terms as Hilbert space, operators, wavelets, multiresolutions, digital image processing, etc. Beginning students, or practitioners from applied areas, might wish to consult the explanations from Section 4.5, in case some terminology might seem unfamiliar. As is the case with all interdisciplinary areas of mathematics and engineering, practitioners from different areas will typically make use of different terminology for what often turns out to be essentially the same concept or the same machinery.

About the Authors

Palle Jorgensen is a Professor at the University of Iowa. He has previously held academic or teaching positions at the University of Pennsylvania, USA; Stanford University, USA; and Aarhus University, Denmark. He has authored more than 300 highly cited research papers and more than 10 books. He has received numerous honors and awards and has delivered many invited lectures, including in 2018 where he was the NSF/CBMS speaker, giving 10 lectures: "Harmonic Analysis: Smooth and Non-smooth," published in volume 128 of the AMS/CBMS book series. His research is interdisciplinary and lies at the crossroads of pure and applied mathematics. Jorgensen is a frequently invited speaker, giving colloquia and conference presentations at universities and centers, both in the US and abroad, most recently at the University of Illinois, USA; the University of Oslo, Norway; the University of Colorado, USA; Rutgers University, USA; the University of Central Florida, USA; Oberwolfach, Germany; Harvard University, USA; Cornell University, USA; Bar-Ilan University, Israel; Ben Gurion University, Israel; and Stockholm University, Sweden. Jorgensen has mentored more than 10 postdocs and has directed more than 30 PhD theses.

Myung-Sin Song is a Professor of Mathematics at Southern Illinois University Edwardsville. Her dissertation was on connecting wavelet image compression with Cuntz-Krieger Algebra. She worked on functional and harmonic analysis of wavelets, the application of wavelet transform on image processing, using computer programming language, and the connection of the engineering of image processing, using wavelet transform and the mathematics of it. Her more recent work is on fractal analysis and its application in image processing, Karhunen–Loève transform (principal component analysis) and spectral theory, lifting scheme, sampling theory and quantization, reproducing kernel Hilbert space and dimension reduction using kernel PCA in machine learning.

Road-map

To facilitate the use of our book by diverse groups of readers, we offer here a multiple-purpose "road map." It suggests how the book can be used in different ways by different reader groups, e.g., in a topics course in a math department, or self-study, or how it could be used by students and readers in neighboring areas, e.g., in engineering, in physics, or by diverse practitioners working with image processing tools. (In physics, the kind of wavelet algorithms we present here is called renormalization.) Our "road map" is a three-part diagram with arrows and chapter numbers. So, suggesting different order of chapters, and paths, for different readers, one arrow diagram for use in a math course, or for self-study, and another chapter selection for engineers, and yet another possible selection of chapters for practitioners from image processing, etc.

Also, readers are referred to Appendix A for basics on operators in Hilbert space, Appendix B for basics on wavelets and multiresolutions, Appendix C for basics on dynamical systems and Cantor dynamics, and Appendix D for basics on Markov chains, generalized wavelet multiresolutions, and representations.

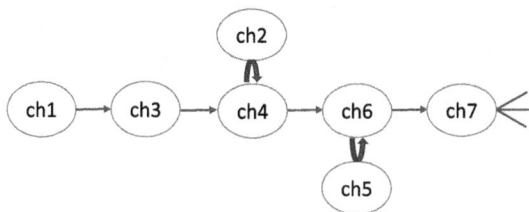

For a math course/seminar, or for self-study.

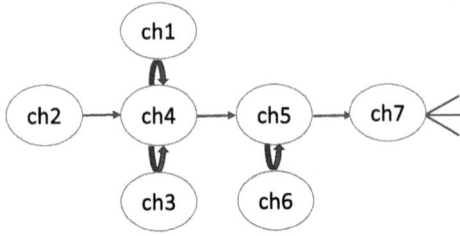

Neighboring areas. Students from engineering, physics, or other.

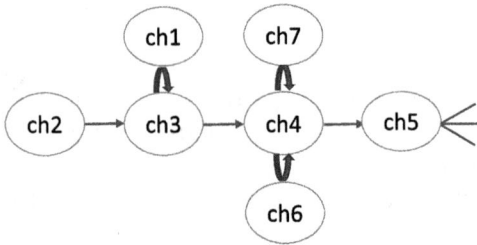

For those involved in digital image processing; pure or applied.

Contents

Chapter 1

Introduction

1.1 Organization of the Book: A Bird's Eye View, and Tips for the Reader

While the chapters in the book follow a logical progression, some readers might want first to pick topics of personal interest. For this reason, we have strived to make each individual chapter reasonably self-contained. Some chapters, and sections, can stand on their own, for example, Appendix C on *Georg Cantor's Chaos*; Chapter 4 on *Wavelet Theory*; Appendix D.2 on *Julia Sets*; and Section 6.7 on *Quantization*.

To further help readers navigate the separate topics, we have included three tables which serve to summarize in detail the main ideas: In Section 3.2, **Glossary**, and in Section 4.8, **List of Names and Discoveries.**

Each chapter builds naturally on what comes before it. Chapter 2 covers Image Compression, and the mathematics of Digital Image Representations. Chapter 3 introduces the key mathematical tool of Multiresolutions; and it leads to the main topic in Chapter 4, Discrete and Continuous Wavelet Transforms, covering both theory and a sample of the many existing applications. Chapter 5 in turn deals with selected tools from dynamics: transfer operators and measures on solenoids. In presenting this, our framework will be general enough to allow realization of the wavelet representations, arising from each of the multi-band wavelet filters from Chapters 3 and 4. Entropy Encoding is the main theme in Chapter 7. In our presentation, we include the key mathematical tools needed for Localization operations, as well as multiple Engineering Applications. In building composite wavelet filters from their atomic constituents, in

Chapters 7 and 8, we introduce the reader to the tool of Factorization and Lifting for Matrix functions.

Our book is based in part on student projects as well as on previous treatments in the literature covering wavelet theory, and its diverse applications to digital image compression, entropy encoding, and various other engineering applications such as lifting scheme, matrix factorization, and multiresolutions with fractals together in this book for readers to understand both the mathematical theory and how algorithmically it works in applications by use of illustrations [Jor03, Son06a, Son06b, JS07, JS14a, JS09, JS10, JS14b, JS18].

Among the motivations for the present book, we stress student projects. While there are many possibilities for such interactive projects, we stress in particular a variety of senior and master's projects for students around the country, aimed at undergraduate and beginning graduate students. In particular, the second named author has already directed such projects (at her university) based on the material in the following Chapters 2, 4, and 5. Specifically, Sections 2.4 and 5.9 are a direct outgrowth of the second named author's experiences with directing senior and master's students projects. These projects have all entailed concrete illustrations, and worked examples, each one making use of "hands-on" digital image compression. Thus, Section 2.4 outlines the key steps involved in image compression, giving in detail construction of associated multi-scale (resolution systems), i.e., in each iteration step at multiple levels, one identifies image decomposition of average, horizontal, verticals, and diagonal details. This further serves to make efficient thresholding as it is used in image compression. See Figures 2.2 to 2.8 for wavelet transform and the resulting compressed images. Also, Section 5.9 outlines in detail those steps of digital image compression which make use of principal component analysis with the dimension reduction method. It further explains how principal components of an image may be computed with eigenvalues and eigenvectors of specific covariance matrix obtained. Then via selection of principal components corresponding to "significant" eigenvalues, the reconstruction of image is performed, which is image compression. See the following Figures 2.13 to 2.25. The student project sections are written in the language of Linear algebra; and thus should be accessible to readers with only a linear algebra background. The projects further serve as introductions to more sophisticated multiresolution constructions which make use of Hilbert space tools.

1.2 Motivation

The notion of multiresolutions (see Appendices B and D) links together such diverse themes as wavelets, image processing, digital algorithms for images; image compression; and the same for speech, radar, etc.; and other related digital signals. And of course, multiresolutions also lie at the heart of analysis on self-similar fractals, both scale self-similarity as well as conformal self-similarity, giving thus Julia sets. The novice might find the following references helpful: [BJ02a, Dau92, Jor05, RW98, SN96]. While the present subsection serves as motivation, our following discussion consists of multiple parts, each one making use of some amount of technical terminology, such as wavelets, image processing, algorithms for digital images, image compression, Fourier series and integrals, multi-scale analyses and multiresolutions, and frequency bands and their representation with the use of closed subspaces in Hilbert space. For the present purpose, our intention is to simply offer a rough road map, and to convey the big picture and some amount of intuitive flavor. As for the technical terms and their use, this will be resumed in detail inside the book. Readers are also referred to the Contents and the Index.

Recently, there has been an enormous amount of interest in the theory and applications of bases in function spaces. The initial impetus was from the discovery of new wavelet tools. But since the applications have come to include a host of interdisciplinary areas, such as the following partial list: signal processing, data compression, turning fingerprints into digital data files, subdivision algorithms for graphics, and the JPEG 2000 encoding of images. As a mathematical subject, wavelet theory involves tools from a host of neighboring fields, functional and harmonic analysis, numerical analysis, mathematics of computation, representation theory, and operator theory.

Constructive basis constructions in function spaces now serve as an alternative to classical Fourier methods, i.e., Fourier series and integrals. The reasons for this is that they are more versatile; in particular, they are better localized, and better adapted to discontinuities; they have a certain form of self-similarity, which makes them suited also for the analysis of fractals and nonlinear dynamical systems. The self-similarity properties of the scaling functions connect them to fractals and nonlinear dynamical systems. This entails a study of multiple scales of a variety of types, including wavelet-multiresolutions.

Such multi-scale analyses and multiresolutions offer fast algorithms. The feature of localization for wavelets is shared by related recursive basis constructions from multiresolutions in Hilbert spaces, for example for fractals and iterated function systems in dynamics. The multiresolutions and locality yield much better pointwise approximations than is possible for traditional Fourier bases.

In signal or image-processing, one is interested in subdividing analogue signals into frequency bands. This idea goes back to Norbert Wiener, but it is of relevance in modern-day wireless signal and image processing.

This suggests a representation theoretic framework. This idea leads to subdivision of analogue signals into frequency bands in signal/image-processing. Motivated by applications to digital filters, we suggest a new representation theoretic framework. We build particular representations creating both Hilbert space $\mathcal{H}$ and algebra representing digital subdivisions, see Appendix A. This leads to a filtered system of closed subspaces in $\mathcal{H}$ such that "non-overlapping frequency bands" correspond to orthogonal subspaces in $\mathcal{H}$; or equivalently to systems of orthogonal projections. Since the different frequency bands must exhaust the signals for the entire system, one looks for orthogonal projections which add to the identity operator in $\mathcal{H}$.

Since time/frequency analysis is non-commutative, one is further faced with a selection of special families of commuting orthogonal projections. When an iteration scheme is applied to the initial generators, one generates new bases and frames by repeated subdivision sequences; that is wavelet families as a recursive scheme.

We hope to bring out both the diversity of these subjects as well as to articulate their unity and interconnections. Developments in one area are likely to inspire advances in another, and a special issue like this is likely to foster cross-fertilization as well as new applications.

Multiresolutions will be the main focus in this book as a multiscale analysis. They offer fast algorithms, and have a host of other applications.

In general, with multiresolutions, one obtains recursive and computational spectral resolutions for multivariable operator systems. They are localized, so better adapted to discontinuities. They offer better numerical schemes.

Multiresolutions are further useful in the study of self-similarity, in the analysis of fractals, and of nonlinear dynamical systems, see Appendices C and D. A special case of this is illustrated by the renormalization property for scaling functions from wavelet theory; and renormalization more generally.

Recursive multiresolutions are basis constructions in Hilbert space, with further applications to analysis of fractals and to iterated function systems in dynamics. Indeed, with multiresolutions, notions of self-similarity and locality, as compared to traditional Fourier bases, yield much better and faster pointwise approximations.

Chapter 2

Wavelet Color Image Compression

This chapter presents how wavelet theory is implemented as algorithms for wavelet digital color image decomposition and reconstruction in image compression. We take various wavelets as concrete examples, discuss their mathematical properties such as orthogonality, vanishing moments, etc., and show the mathematical algorithm for wavelet image decomposition and reconstruction with image matrix, and finally present the MATLAB wavelet image compression results [Son06a, Son06b, MMOP02] (Figure 2.1).

2.1 Introduction

Wavelets are functions which allow data analysis of signals or images according to scales or resolutions. The processing of signals by wavelet algorithms in fact works much the same way the human eye does; or the way a digital camera processes visual scales of resolutions, and intermediate details. But the same principle also captures cell phone signals, and even digitized color images used in medicine. Wavelets are of real use in these areas, for example in approximating data with sharp discontinuities such as choppy signals, or pictures with lots of edges. The novice might find the following references helpful: [DJ05, GWE04, Mar82, Son06b, Wic94].

While wavelets form perhaps a chapter in function theory, we show that the algorithms that result are key to the processing of numbers, or more precisely of digitized information, signals, time series, still-images, movies, color images, etc. Thus, applications of the wavelet idea include big parts of signal and image processing, data compression, fingerprint encoding, and many other fields of science and engineering. This thesis focuses on

Fig. 2.1 The second level wavelet decomposition of a digital image; see the following discussion.

the processing of color images with the use of custom-designed wavelet algorithms and mathematical threshold filters.

Although there have been a number of recent papers on the operator theory of wavelets, there is a need for a tutorial which explains some applied trends from scratch to operator theorists. Wavelets as a subject is highly interdisciplinary and it draws in crucial ways on ideas from the outside world. We aim to outline various connections between Hilbert space geometry and image processing. Thus, we hope to help students and researchers from one area understand what is going on in the other. One difficulty with communicating across areas is a vast difference in lingo, jargon, and mathematical terminology.

With hands-on experiments, our book is meant to help create a better understanding of links between the two sides, math and images. It is a delicate balance deciding what to include. In choosing, we had in mind students in operator theory, stressing explanations that are not easy to find in the journal literature.

Our book results extend what was previously known, and we hope yield new insight into scaling and of representation of color images; we have especially aimed for better algorithms.

This chapter concludes with a set of computer-generated images which serve to illustrate our ideas and our algorithms, and also with the resulting compressed images.

2.1.1 *Overview*

How wavelets work in image processing is analogous to how our eyes work. Depending on the location of the observation, one may perceive a forest differently. If the forest were observed from the top of a skyscraper, it will be observed as a blob of green; if it were observed from a moving car, it will be observed as the trees in the forest flashing past, thus the trees are now recognized. Nonetheless, if it is observed by one who actually walks around it, then more details of the trees such as leaves and branches, and perhaps even the monkey on the top of the coconut tree, may be observed. Furthermore, pulling out a magnifying glass may even make it possible to observe the texture of the trees and other little details that cannot be perceived by bare human eyes [Mac01, Mar82].

Wavelet Image Processing enables computers to store an image in many scales of resolutions, thus decomposing an image into various levels and types of details and approximations with different-valued resolutions. Hence, making it possible to zoom in to obtain more details of the trees, leaves, and even a monkey on top of the tree. Wavelets allow one to compress the image using less storage space with more details of the image.

The advantage of decomposing images to approximate and detail parts as in Section 2.3.3 is that it enables to isolate and manipulate the data with specific properties. With this, it is possible to determine whether to preserve more specific details. For instance, keeping more vertical details instead of keeping all the horizontal, diagonal, and vertical details of an image that has more vertical aspects. This would allow the image to lose a certain amount of horizontal and diagonal details, but would not affect the image in human perception.

As mathematically illustrated in Section 2.3.3, an image can be decomposed into approximate horizontal, vertical, and diagonal details. N levels of decomposition are done. After that, quantization is done on the decomposed image where different quantization maybe done on different components thus maximizing the amount of needed details and ignoring "not-so-wanted" details. This is done by thresholding where some coefficient values for pixels in images are "thrown out" or set to zero or some "smoothing" effect is done on the image matrix. This process is used in JPEG2000.

2.1.2 *Motivation*

In many papers and books, the topics in wavelets and image processing are discussed mostly in one extreme, namely in terms of engineering aspects of it or wavelets are discussed in terms of operators without it being specifically mentioned how it is used in its application in engineering. In this book, the authors stress interplay between theory and applications. We build in part on [SCE01, Use01], and [Vet01], but with more pedagogical insights added about mathematical properties such as properties from Operator Theory, Functional Analysis, etc. of wavelets playing a major role in results in wavelet image compression. Our book aims in establishing, if not already established, or improving the connection between the mathematical aspects of wavelets and their application in image processing. Also, our book discusses on how the images are implemented with computer programs, and how wavelet decomposition is done on the digital images in terms of computer programs, and in terms of mathematics, in the hope that the communication between mathematics and engineering will improve, thus bringing greater benefits to mathematicians and engineers.

2.2 Wavelet Color Image Compression

2.2.1 *Methods*

A graphic organization of the method is summarized in Figure 2.9, outlining sequential steps in the wavelet image compression process. The whole process of wavelet image compression (see, e.g., Figures 2.2 to 2.8 for wavelet transform, and the resulting compressed images, in Figures 2.13 to 2.25) is performed as follows: An input image is taken by the computer, forward wavelet transform is performed on the digital image, thresholding is done on the digital image, entropy coding is done on the image where necessary, thus the compression of the image is done on the computer. Then with the compressed image, reconstruction of the wavelet transformed image is done, after which inverse wavelet transform is performed on the image, thus the image is reconstructed. In some cases, zero-tree algorithm [Sha93b] is used, which it is known to give better compression, but it was not implemented here.

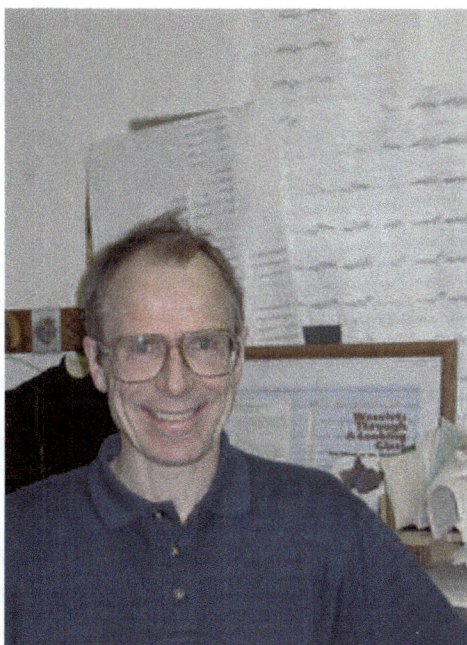

Fig. 2.2 The original image before the wavelet decomposition.

2.2.1.1 *Forward wavelet transform*

Various wavelet transforms are used in this step. Namely, Daubechies wavelets, Coiflets, biorthogonal wavelets, and Symlets. These various transforms are implemented to observe how various mathematical properties such as symmetry, number of vanishing moments, and orthogonality affect the result of the compressed image. Advantages of a short support is that it preserves locality. The Daubechies wavelets used are orthogonal, so are the Coiflets. Symlets have the property of being close to symmetric. The biorthogonal wavelets are not orthogonal, but not having to be orthogonal gives more options to a variety of filters, such as symmetric filters, thus allowing them to possess the symmetric property.

MATLAB has a subroutine called wavedec2 which performs the decomposition of the image for you up to the given desired level (N) with the given

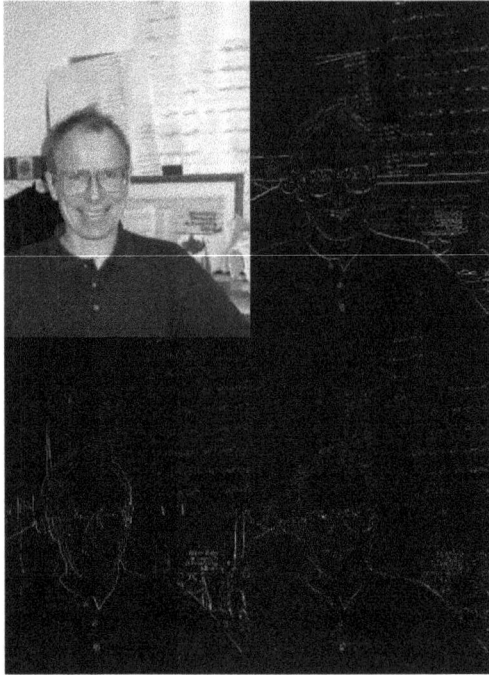

Fig. 2.3   Wavelet decomposition of an image component — 1st level decomposition.

desired wavelet ("wname"). Since there are three components to deal with, the wavelet transform was applied componentwise. "wavedec" is a two-dimensional wavelet analysis function. $[C,S]$ = wavedec2(X,N, "wname") returns the wavelet decomposition of the matrix X at level N, using the wavelet named in string "wname". Outputs are the decomposition vector C and the corresponding bookkeeping matrix S [MMOP02]. Here, the image is taken as the matrix X.

2.2.1.2 *Thresholding*

Since the whole purpose of this project was to compare the performance of each image compression using different wavelets, fixed thresholds were used.

Soft threshold was used in this project in the hope that the drastic differences in gradient in the image would be noted less apparently. The

Fig. 2.4 Wavelet decomposition of an image component — 2nd level decomposition.

Fig. 2.5 The original image before the wavelet decomposition.

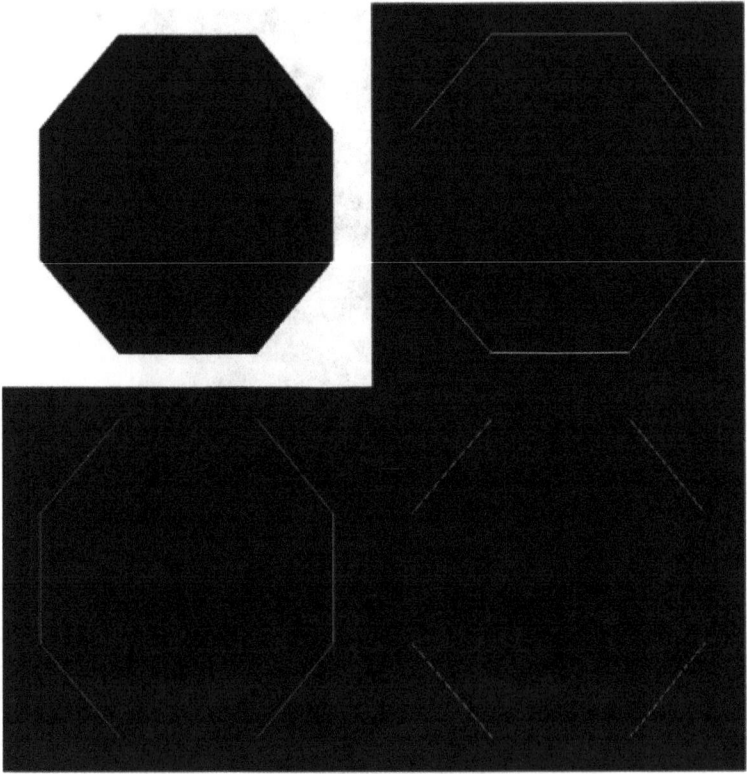

Fig. 2.6   Wavelet decomposition of an image component — 1st level decomposition.

soft and hard thresholdings T_{soft}, T_{hard} are defined as follows:

$$T_{\text{soft}}(x) = \begin{cases} 0 & \text{if } |x| \leq \lambda \\ x - \lambda & \text{if } x > \lambda \\ x + \lambda & \text{if } x < -\lambda \end{cases} \qquad (2.1)$$

$$T_{\text{hard}}(x) = \begin{cases} 0 & \text{if } |x| \leq \lambda \\ x & \text{if } |x| > \lambda \end{cases}, \qquad (2.2)$$

where $\lambda \in \mathbb{R}_+$. As it could be observed by looking at the definitions, the difference between them is related to how the coefficients larger than a threshold value λ in absolute values are handled. In hard thresholding, these coefficient values are left alone. Unlike in hard thresholding, the coefficient values area decreased by λ if positive and increased by λ if negative [Wal02].

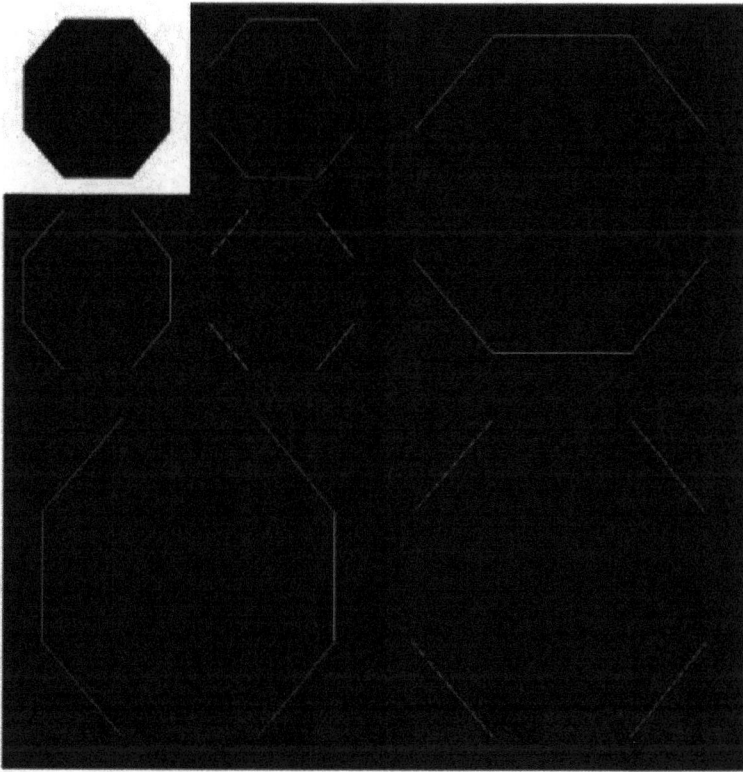

Fig. 2.7 Wavelet decomposition of an image component — 2nd level decomposition.

MATLAB has this subroutine called wthrmngr, which computes the global threshold or level-dependent thresholds depending on the option and method. The options available are global threshold and level-dependent threshold, and the global threshold is used in the program. However, fixed threshold values were used so as to have the same given condition for every wavelet transform to compare the performances of different conditions. Here, fixed thresholds 10 and 20 were used. For the lossless compression, 0 is used as the threshold for an obvious reason.

2.2.1.3 *Entropy encoding*

Entropy is defined as

$$H(s) = -\sum_{i=1}^{q} P(s_i) \log_2(P(s_i)),$$

Fig. 2.8 Wavelet decomposition of an image component — 3rd level decomposition.

where s_i are codewords and s is the message. Entropy coding uses codewords with varying lengths, here codewords with short lengths are used for values that have to be encoded more often, and the longer codewords are assigned for less encoded values. $H(s)$ measures the amount of information in the message, i.e., the minimal number of bits needed to encode one word of the message. Unfortunately, the entropy encoding was not implemented on the codes for the color image compression using wavelets. However, Shannon entropy, which is defined in what follows, was used in the code for the image compression with wavelet packets [Son06a, Son06b]. See [BJ02a].

The Shannon entropy functional is defined by

$$M(c\{b_j\}) = -\sum_{n=1} M|\langle c, b_j \rangle|^2 \log |\langle c, b_j \rangle|^2. \tag{2.3}$$

Also, entropy could be viewed as a quantity that measures the amount of uncertainty in a probability distribution, or equivalently of the amount of information obtained from one sample from the probability space.

2.2.1.4 *Reconstruction of wavelet transformed image*

At this step, the significance map is taken and with the amplitudes of the non-zero valued wavelet coefficients, the wavelet transformed image is reconstructed.

2.2.1.5 *Inverse wavelet transformation*

The wavelet parameters are converted back into an image almost identical to the original image. How identical they are will be dependent upon whether the compression was lossy or lossless.

2.2.2 Wavelets

Compactly supported wavelets are functions defined over a finite interval and having an average value of zero. The basic idea of the wavelet transform is to represent any arbitrary function $f(x)$ as a superposition of a set of such wavelets or basis functions. These basis functions are obtained from a single prototype wavelet called the mother wavelet $\psi(x)$, by dilations or scaling and translations. Wavelet bases are very good at efficiently representing functions that are smooth except for a small set of discontinuities.

For each $n, k \in \mathbb{Z}$, define $\psi_{n,k}(x)$ by

$$\psi_{n,k}(x) = 2^{n/2}\psi(2^n x - k). \tag{2.4}$$

Construct the function $\psi(x)$, L^2 on $\mathbb{R}$, such that $\{\psi_{n,k}(x)\}_{n,k \in \mathbb{Z}}$ is an orthonormal basis on $\mathbb{R}$. As mentioned before, $\psi(x)$ is a wavelet and the collection $\{\psi_{n,k}(x)\}_{n,k \in \mathbb{Z}}$ is a wavelet orthonormal basis on $\mathbb{R}$; this framework for constructing wavelets involves the concept of a multiresolution analysis or MRA.

2.2.2.1 *Multiresolution analysis*

Multiresolution analysis is a device for computation of basis coefficients in $L^2(\mathbb{R}) : f = \sum \sum c_{n,k}\psi_{n,k}$. It is defined as follows. See [Kei04]. Define

$$V_n = \{f(x)|f(x) = 2^{n/2}g(2^n x), g(x) \in V_0\},$$

where

$$f(x) = \sum_{n \in \mathbb{Z}} \langle f, \phi(\cdot - n) \rangle \phi(x - n).$$

Then a multiresolution analysis on $\mathbb{R}$ is a sequence of subspaces $\{V_n\}_{n \in \mathbb{Z}}$ of functions L^2 on $\mathbb{R}$, satisfying the following properties:

(a) For all $n, k \in \mathbb{Z}, V_n \subseteq V_{n+1}$.
(b) If $f(x)$ is C_c^0 on $\mathbb{R}$, then $f(x) \in \overline{span}\{V_n\}_{n \in \mathbb{Z}}$. That is, given $\epsilon > 0$, there is an $n \in \mathbb{Z}$ and a function $g(x) \in V_n$ such that $\|f - g\|_2 < \epsilon$.
(c) $\bigcap_{n \in \mathbb{Z}} V_n = \{0\}$.
(d) A function $f(x) \in V_0$ if and only if $2^{n/2} f(2^n x) \in V_n$.
(e) There exists a function $\phi(x)$, L^2 on $\mathbb{R}$, called the scaling function such that the collection $\phi(x - n)$ is an orthonormal system of translates and $V_0 = \overline{span}\{\phi(x - n)\}$.

Definition 2.1. Let $\{V_J\}$ be an MRA with scaling function $\phi(x)$, which satisfies (2.14), and scaling filter $h(k)$, where $h(k) = \langle 2^{-1/2} \phi(\frac{x}{2}), \phi(x - k) \rangle$. Then the wavelet filter $g(k)$ is defined by

$$g(k) = (-1)^k \overline{h(1 - k)}$$

and the wavelet, by

$$\psi(x) = \sum_{k \in \mathbb{Z}} g(k) \sqrt{2} \phi(2x - k).$$

See [Kei04].

Then $\{\psi_{n,k}(x)\}_{n,k \in \mathbb{Z}}$ is a wavelet orthonormal basis on $\mathbb{R}$.

Definition 2.2. The orthogonal projection of an arbitrary function $f \in L^2$ onto V_n is given by

$$P_n f = \sum_{k \in \mathbb{Z}} \langle f, \phi_{n,k} \rangle \phi_{n,k}.$$

[Kei04]. As k varies, the basis functions $\phi_{n,k}$ are shifted in steps of 2^{-n}, so $P_n f$ cannot represent any detail on a scale smaller than that. We say that the functions in V_n have the resolution 2^{-n} or scale 2^{-n}. Here, $P_n f$ is called an approximation to f at resolution 2^{-n}. For a given function f, an MRA provides a sequence of approximations $P_n f$ of increasing accuracy [Kei04]. We include the following known proof for the benefit of the readers.

Theorem 2.3. *[Wal02] For all $f(x) \in C_c^0(\mathbb{R})$, $\lim_{n \to \infty} \|P_n f - f\|_2 = 0$.*

Proof. Let $\epsilon > 0$. Then there exists $N \in \mathbb{Z}$ and $g(x) \in V_j$ such that $\|f - g\|_2 < \epsilon/2$. By 2.1, $g(x) \in V_n$ and $P_n g(x) = g(x)$ for all $n \leq N$. Thus,

$$\|f - P_n f\|_2 = \|f - g + P_n g - P_n f\|$$

$$\leq \|f - g\|_2 + \|P_n(f - g)\|_2$$

$$\leq 2\|f - g\|_2$$

$$< \epsilon.$$

where Minkowski's and Bessel's inequalities are applied. Since this holds for all $n \geq N$, the proof is complete. $\qquad\square$

Definition 2.4. The difference between the approximations at resolution 2^{-n} and 2^{-n-1} is called the fine detail at resolution 2^{-n}, which is as follows:

$$Q_n f(x) = P_{n+1} f(x) - P_n f(x)$$

or

$$Q_n f = \sum_{k \in \mathbb{Z}} \langle f, \psi_{n,k} \rangle \psi_{n,k}.$$

Q_n is also an orthogonal projection and its range W_n is orthogonal to V_n, where the following holds:

$$V_n = \{f | P_n f = f\} \tag{2.5}$$

$$W_n = \{f | Q_n f = f\} \tag{2.6}$$

$$V_n \oplus W_n = V_{n+1} \tag{2.7}$$

$$\psi \in V_{-1} \ominus V_0 = \{f | f \in V_{-1}, f \perp V_0\} = W_0. \tag{2.8}$$

Theorem 2.5 ([Dau92, Wal02]). *There are choices of the numbers h and g in 2.1 such that $\{\psi_{n,k}(x)\}_{n,k \in \mathbb{Z}}$ is a wavelet orthonormal basis on $\mathbb{R}$.*

Proof. We must show orthonormality and completeness. As for completeness, we have

$$\bigcap_{n \in \mathbb{Z}} V_n = 0 \tag{2.9}$$

and

$$\overline{\bigcup_n V_n} = L^2(\mathbb{R}). \tag{2.10}$$

See Figures 7.1, 4.2.1, 6.22 for the description of these $\{V_n\}$ spaces. Then we have $\{\psi_{n,k} | k \in \mathbb{Z}\} = W_n = V_n \ominus V_{n-1}$. Hence, $\{\psi_{n,k}\}_{n,k \in \mathbb{Z}}$ is complete if and only if $\sum_n W_n = L^2(\mathbb{R})$ hold, and this is true if and only if (2.9) and (2.10) hold. Since we have those conditions for $\{\psi_{n,k}\}_{n,k \in \mathbb{Z}}$, it is complete.

Now, as for the orthonormality,

$$\langle \psi_{n,k}, \psi_{n,l} \rangle = \langle 2^{n/2}\psi(2^n x - k), 2^{n/2}\psi(2^n x - l) \rangle$$
$$= \langle \psi_{0,k}, \psi_{0,l} \rangle = \delta(k - l).$$

To prove orthonormality between scales, let $n, n' \in \mathbb{Z}$ with $n' < n$, and let $k, k' \in \mathbb{Z}$ be arbitrary. Since $\psi(x) \in V_1$, $\psi_{0,k'}(x) \in V_1$ also. Then we have $\psi_{n',k'} \in V_{n'+1}$. Since $\langle \psi_{0,k}, \psi_{0,l} \rangle = 0$ for all $k, l \in \mathbb{Z}$, it follows that $\langle \psi_{n,k}, \phi_{n,l} \rangle = 0$, for all $n, k, l \in \mathbb{Z}$. Given $f(x) \in V_n$ we know that $f(x) = \sum_k \langle f, \phi_{n,k} \rangle \phi_{n,k}(x)$. Hence, for $f(x) \in V_n$,

$$\langle \psi_{n,l}, f \rangle = \langle \psi_{n,l}, \sum_k \langle f, \phi_{n,k} \rangle \phi_{n,k} \rangle$$
$$= \sum_k \overline{\langle f, \phi_{n,k} \rangle} \langle \psi_{n,l}, \phi_{n,k} \rangle = 0.$$

Since $n' < n$, $V_{n'+1} \subseteq V_n$ and since $\psi_{n',k'} \in V_{n'+1}$, $\psi_{n',k'} \in V_n$ also. Hence, $\langle \psi_{n,k}, \psi_{n',k'} \rangle = 0$. Therefore, $\{\psi_{n,k}(x)\}_{n,k \in \mathbb{Z}}$ is a wavelet orthonormal basis on $\mathbb{R}$. See [Wal02].   □

2.2.2.2 *Symmetry*

Symmetric filters are preferred for they are most valuable for minimizing the edge effects in the wavelet representation of discrete wavelet transform (DWT) of a function; large coefficients resulting from false edges due to periodization can be avoided.

Since orthogonal filters in exception to Haar filter cannot be symmetric, biorthogonal filters are almost always selected for image compression application [Wal02].

2.2.2.3 *Vanishing moments*

Vanishing moments are defined as follows: From the definition of multiresolution analysis (MRA), any wavelet $\psi(x)$ that comes from an MRA must satisfy

$$\int_{\mathbb{R}} \psi(x)dx = 0. \qquad (2.11)$$

The integral (2.11) is referred to as the zeroth moment of $\psi(x)$, so that if (2.11) holds, we say that $\psi(x)$ has its zeroth moment vanishing. The integral

$\int_{\mathbb{R}} x^k \psi(x) dx$ is referred to as the kth moment of $\psi(x)$ and if $\int_{\mathbb{R}} x^k \psi(x)$ $dx = 0$, we say that $\psi(x)$ has its kth moment vanishing [Wal02].

We may encounter a situation where we have different number of vanishing moments on the analysis filters than on the reconstruction filters. As a matter of this fact, it is possible to have different number of vanishing moments on the analysis filters than on the reconstruction filters. Vanishing moments on the analysis filters are desired for small coefficients in the transform as a result, whereas vanishing moments on the reconstruction filter result in fewer blocking artifacts in the compressed image, thus the latter is desired. Thus, having sufficient vanishing moments, which maybe different in numbers on each filter, are advantageous.

2.2.2.4 *Size of the filters*

Long analysis filters result in greater computation time for the wavelet or wavelet packet transform. Long reconstruction filters can create unpleasant artifacts in the compressed image for the following reason. The reconstructed image is made up of the superposition of only a few scaled and shifted reconstruction filters. So, features of the reconstruction filters, such as oscillations or lack of smoothness, can be obviously noted in the reconstructed image. Smoothness can be guaranteed by requiring a large number of vanishing moments in the reconstruction filter. However, such filters tend to be oscillatory [Wal02]. Also see [Vet01].

2.2.3 *Various wavelets*

For later use in computations, we recall here the specific wavelets needed:

2.2.3.1 *Haar Wavelet*

Haar wavelet is the only known wavelet that is compactly supported, orthogonal, and symmetric [Res05].

The Haar system is defined as follows in [Wal02]. Define

$$I_{n,k} = [2^{-n}k, 2^{-n}(k+1))].$$

Let $p(x) = \chi_{[0,1)}(x)$, and for each $n, k \in \mathbb{Z}$, define $p_{n,k}(x) = 2^{n/2}p(2^n x - k)$. The collection $\{\phi_{n,k}(x)\}_{n,k \in \mathbb{Z}}$ is referred to as the system of Haar scaling functions.

(a) For each $n, k \in \mathbb{Z}$, $p_{n,k}(x) = 2^{n/2}\chi_{I_{n,k}}(x)$, so that $p_{n,k}(x)$ is supported on the interval $I_{n,k}$ and does not vanish on that interval. Therefore, we refer to the scaling function $p_{n,k}(x)$ as being associated with the interval $I_{n,k}$.

(b) For each $n, k \in \mathbb{Z}$, $\int_{\mathbb{R}} p_{n,k}(x)dx = \int_{I_{n,k}} p_{n,k}(x)dx = 2^{n/2}$ and

$$\int_{\mathbb{R}} |p_{n,k}(x)|^2 dx = \int_{I_{n,k}} |p_{n,k}(x)|^2 dx = 1.$$

Definition 2.6. Let $h(x) = \chi_{[0,1/2)}(x) - \chi_{[1/2,1)}(x)$, and for each $n, k \in \mathbb{Z}$ define $h_{n,k}(x) = 2^{n/2}h(2^n x - k)$. The collection $\{h_{n,k}(x)\}_{n,k\in\mathbb{Z}}$ is referred to as the Haar system on $\mathbb{R}$. For each $n \in \mathbb{Z}$, the collection $\{h_{n,k}(x)\}_{n,k\in\mathbb{Z}}$ is referred to as the system of scale j Haar functions.

Definition 2.7. Given $J, N \in \mathbb{N}$ with $J < N$ and a finite sequence $c_0 = \{c_0\}_{k=0}^{2^N-1}$, the discrete Haar transform of c_0 is defined by $\{d_j(k)|1 \leq j \leq J; 0 \leq k \leq 2^{N-j} - 1\} \cup \{c_J(k)\|0 \leq k \leq 2^{N-j} - 1\}$, where

$$c_j(k) = \frac{1}{\sqrt{2}}c_{j-1}(2k) + \frac{1}{\sqrt{2}}c_{j-1}(2k+1)d_j(k)$$

$$= \frac{1}{\sqrt{2}}c_{j-1}(2k) + \frac{1}{\sqrt{2}}c_{j-1}(2k+1) \tag{2.12}$$

$$c_{j-1}(2k) = \frac{1}{\sqrt{2}}c_j(k) + \frac{1}{\sqrt{2}}d_j(k)c_{j-1}(2k+1)$$

$$= \frac{1}{\sqrt{2}}c_j(k) - \frac{1}{\sqrt{2}}d_j(k). \tag{2.13}$$

Haar wavelets are basically the same as Daubechies wavelets db1 (in MATLAB) or Daub4. Haar wavelets are examples of compactly supported wavelets. The compact support of the Haar wavelets enables the Haar decomposition to have a good time localization. Specifically, this means that the Haar coefficients are effective for locating jump discontinuities and also for the efficient representation of signals with small support. However, the fact is that they have jump discontinuities (Section 5.4.3), in particular in the poorly decaying Haar coefficients of smooth functions and in the blockiness of images reconstructed from subsets of the Haar coefficients. [Wal02].

2.2.3.2 *Daubechies wavelets constructions*

In order to construct compactly supported, orthogonal wavelets, we first look at the dilation equation

$$\phi(x) = \sqrt{2}\sum_{k\in\mathbb{Z}} h(k)\phi(2x - k) \tag{2.14}$$

and the wavelet equation

$$\psi(x) = \sqrt{2} \sum_{k \in \mathbb{Z}} g(k)\phi(2x - k). \tag{2.15}$$

Note from these two equations that the compactness of the support of ϕ and ψ can be achieved if we choose the number of nonvanishing coefficients $\{h(k)\}$, that is, the filter length, to be finite. This implies that $m_0(\omega) = \sum_j h(j)e^{-in\omega}/\sqrt{2}$ is a trigonometric polynomial.

One can see that once we know ϕ at the integers, the values of ϕ at the dyadic points $k/2^n$ then can be obtained recursively using the scaling equation. Once we find ϕ, we can use (2.15) to generate ψ.

So Daubechies's approach to finding ϕ and ψ is to first determine the finite number of filter coefficients $h(j)$ such that orthogonality and smoothness or moment conditions are guaranteed. To find $h(j)$, we start from the Fourier domain where the orthogonality condition for the scaling function ϕ is

$$|m_0(\omega)|^2 + |m_0(\omega + \pi)|^2 = 1. \tag{2.16}$$

The condition that the first N vanishing moments is

$$\int_{\mathbb{R}} x^k \psi(x)dx = 0 \quad \text{for } k = 0, 1, \ldots, N - 1. \tag{2.17}$$

To satisfy the moment condition (2.17), $m_0(\omega)$ has to assume the following form:

$$m_0(\omega) \propto \left(\frac{1 + e^{i\omega}}{2}\right)^N. \tag{2.18}$$

Now, define

$$M_0(\omega) = |m_0(\omega)|^2,$$

where $M_0(\omega)$ is a polynomial in $\cos(\omega)$, the moment-vanishing condition implies

$$M_0(\omega) = (\cos^2 \omega/2)^N L(\omega), \tag{2.19}$$

where $L(\omega)$ is a polynomial in $\cos(\omega)$, and the orthogonality condition gives

$$M_0(\omega) + M_0(\omega + \pi) = 1. \tag{2.20}$$

By the half-angle identity, we can write $L(\omega) = P(\sin^2 \omega/2)$. Now, it is left to us to find the form of the polynomial P, "take its square root" to get $m_0(\omega)$, and identify the coefficients $\{h(j)\}$. Let $y = \sin^2 \omega/2$, using (2.19) and (2.20), we see that P satisfies

$$(1-y)^N P(y) + y^N P(1-y) = 1, \tag{2.21}$$

or

$$P(y) = (1-y)^{-N}(1 - y^N P(1-y)). \tag{2.22}$$

It turns out that the lowest-degree polynomial that satisfies (2.21) is $N-1$. So we can find the form of the polynomial P with degree $N-1$ explicitly from (2.22) by expanding $(1-y)^{-N}$ in a Taylor series and retaining terms up to order $N-1$:

$$P(y) = (1-y)^{-N}(1 - y^N P(1-y)) = \sum_{k=0}^{N-1} \binom{N+k-1}{k} y^k. \tag{2.23}$$

Note that $P(y) \geq 0$ for $y \in [0, 1]$.

To Summarize what we have done, we have established

$$|m_0(\omega)|^2 = M_0(\omega) = (\cos^2 \omega/2)^N P(\sin^2 \omega/2), \tag{2.24}$$

where P is given by (2.23), such that $m_0(\omega)$ satisfies the required moment-vanishing condition and the orthogonality condition. Now is it left to us to "take the square root" of P to obtain $m_0(\omega)$ and it will be done by spectral factorization.

We write,

$$A(\omega) = P\left(\frac{1-\cos(\omega)}{2}\right) = \alpha \prod_{j=0}^{N-1} (\cos(\omega) - c_j),$$

where we have regarded P as a polynomial of degree $N-1$, and expressed it in terms of its roots $\{c_j\}$ (α is a constant). Since the polynomial has real coefficients, the roots c_j either are real or occur in complex conjugate pairs.

On the other hand, we can also write the original polynomial P in terms of $z = e^{i\omega}$,

$$A(\omega) = \alpha \prod_{j=0}^{N-1} \left(\frac{z+z^{-1}}{2} - c_j\right) = P\left(\frac{1 - (z+z^{-1}/2)}{2}\right) = P(z). \tag{2.25}$$

The zeros of $p(z)$ appear in quadruplets $\{z_j, \bar{z}_j, z_j^{-1}, \bar{z}_j^{-1}\}$ if $z_j \in \mathbb{C}$, and in doublets $\{r_j, r_j^{-1}\}$ if $z_j = r_j \in \mathbb{R}$. So, we can write

$$p(z) = \alpha z^{-(N-1)} \prod_{j=0}^{N-1} \left(\frac{z^2}{2} - c_j z + \frac{1}{2}\right)$$

$$= \alpha' z^{-(N-1)} \prod_{j} (z - z_j)(z - \bar{z}_j)(z - z_j^{-1})(z - \bar{z}_j^{-1})$$

$$\times \prod_{k} (z - z_k)^2 ((z - \bar{z}_k)^2 \prod_{l} (z - r_l)(z - r_l^{-1}).$$

Earlier, we have separated the case $z_j = e^{i\alpha j}$, where $z_j = \bar{z}_j^{-1}$ and the quadruplet reduces to a doublet of degeneracy 2 [Res05].

The Daubechies wavelets are orthogonal wavelets, which are energy or norm preserving. There are a number of Daubechies wavelets, Daub$J$, where $J = 4, 6, \ldots, 20$. The easiest way to understand this transform is just to treat them as simple generations of the Daub4 transform with the scaling and translation factors. The most apparent difference between each of them is the length of the supports of their scaling signals and wavelets. Daub4 wavelet is the same as the Haar wavelet. Daub4 wavelet preserves the energy due to its orthogonality and the proof of this could be found on pp. 39–40 of [Wal99a]. Daub4 transform is suitable for identifying features of the signal that are related to turning points in its graph [Wal99a]. Now, one might wonder why we have so many different Daub$J$s and what their advantages and disadvantages are. Daub6 often produces smaller size fluctuation values than those produced by Daub4 transform. The types of signals for which this occurs are the ones that are obtained from the sample of analog signals that are at least three times continuously differentiable. These kinds of signals are approximated better, over a large proportion of their values, by quadratic approximations. The curve graphs of quadratic functions enable them to provide superior approximations to the parts of the signal that are near to the turning points in the graphs. So for, signal compression Daub6 transform generally does a better job. However, the fact is that Daub4 is better in approximating signals better approximated by linear approximation [Wal99a].

If $H = H'$ and $G = G'$ in a biorthogonal set [Dau92] of quadrature filters, then the pair H, G is called an orthogonal quadrature filter pair which is a pair of operators and is defined as follows:

$$Hu(i) = \sum_{j=-\infty}^{\infty} h(2i - j)u(j), \quad i \in \mathbb{Z} \qquad (2.26)$$

$$Gu(i) = \sum_{j=-\infty}^{\infty} g(2i - j)u(j), \quad i \in \mathbb{Z}. \tag{2.27}$$

In addition, the following conditions hold:

- Self-duality: $H'H^* = G'G^* = I$
- Independence: $GH^* = HG^* = 0$
- Exact reconstruction: $H^*H + G^*G = I$
- Normalization: $H1 = \sqrt{2}1$

H is the low-pass filter and G is the high-pass filter.

The first two conditions may be expressed in terms of the filter sequences h, g, which respectively define H, G:

$$\sum_k h(k)\bar{h}(k + 2n) = \delta(n) = \sum_k g(k)\bar{g}(k + 2n) \tag{2.28}$$

$$\sum_k g(k)\bar{h}(k + 2n) = 0 = \sum_k h(k)\bar{g}(k + 2n). \tag{2.29}$$

See [Wic94].

2.2.3.3 *Coiflets*

Coiflets are designed so as to maintain a close match between the trend values and the original signal values. All of the coiflets, Coif$I$, $I = 6, 12, 18, 24, 30$ are defined in a similar way as Daubechies wavelets, but they have some different properties. Coif6 transform produces a much closer match between trend subsignals and the original signal values than the match that any of the Daub$J$ transforms can produce. This means that the Coif$I$ wavelets have nearly symmetric graphs [Wal99a].

Coifman wavelet systems are similar to Daubechies wavelet systems (in rank 2) in that they have a maximal number of vanishing moments, but the vanishing of moments are equally distributed between the scaling function and the wavelet function. In contrast to the case for Daubechies wavelets, there is no formula for Coiflets of arbitrary genus, and there is no formal proof of their existence for arbitrary genus at this time. There are numerical solutions using Newton's method which work well until round-off error gives problems, up to about genus 20 (round-off error is also a problem in calculating the Daubechies scaling vector numerically beyond this same range with spectral factorization, even though the formulas are valid and give an existence theorem for every genus) [RW98].

If we used Daubechies wavelets in the same way, one cannot get the same approximation results, except to low order.

It is very advantageous to have a high number of vanishing moments for ψ; it leads to high compressibility because the fine scale wavelets coefficients of a function would be essentially zero where the function is smooth. Since $\int_{\mathbb{R}} \phi(x) = 1$, the same thing can never happen for the $\langle f, \phi_{n,k} \rangle$. Nevertheless, if $\int_{\mathbb{R}} x^l \phi(x) dx = 0$ for $l = 1, \ldots, L$, then we can apply the same Taylor expansion argument and conclude that for N large, $\langle f, \phi_{-N,k} \rangle \approx 2^{N/2} f(2^{-N} k)$, with an error that is negligibly small where f is smooth. This means that we have a remarkably simple quadrature rule to go from the sample of f to its fine scale coefficients $\langle f, \phi_{-N,k} \rangle$. For this reason, R. Coifman suggested that it might be worthwhile to construct orthonormal wavelet bases with vanishing moments not only for ψ, but also for ϕ. See [Dau92].

2.2.3.4 *Biorthogonal wavelets*

The biorthogonal wavelets have bases that are defined in a way that has weaker definition of the orthogonal wavelet bases. Though the orthogonal wavelet's filter has self-duality only, the biorthogonal wavelet's filter has duality. Since the orthogonality of the filter makes the wavelet energy-preserving as proven in [Wal99a], the biorthogonal wavelets are not energy-preserving.

Current compression systems use biorthogonal wavelets instead of orthogonal wavelets, despite the fact that they are not energy-preserving. The fact that biorthogonal wavelets are not energy-preserving is not a big problem since there are linear phase biorthogonal filter coefficients, which are "close" to being orthogonal [Use01]. The main advantage of the biorthogonal wavelet transform is that it permits the use of a much broader class of filters, and this class includes the symmetric filters. The biorthogonal wavelet transform is advantageous because it can use linear phase filters which give symmetric outputs when presented with symmetric input. This transform is called the symmetric wavelet transform and it solves the problems of coefficient expansion and border discontinuities. See [Use01].

A quadruplet H, H', G, G' of convolution operators or filters is said to form a set of biorthogonal quadrature filters if the filters satisfy the following conditions:

- Duality: $H'H^* = G'G^* = I = HH'^* = GG'^*$
- Independence: $G'H^* = H'G^* = 0 = GH'^* = HG'^*$

- Exact reconstruction: $H^*H' + G^*G' = I = H'^*H + G'^*G$
- Normalization: $H1 = H'1 = \sqrt{2}1$ and $G1 = G'1 = 0$

H and H' are the low-pass filters and G and G' are the high-pass filters.

The first two conditions may be expressed in terms of the filter sequences h, h', g, g', which, respectively, define H, H', G, G':

$$\sum_k h'(k)\bar{h}(k + 2n) = \delta(n) = \sum_k g'(k)\bar{g}(k + 2n) \tag{2.30}$$

$$\sum_k g'(k)\bar{h}(k + 2n) = 0 = \sum_k h'(k)\bar{g}(k + 2n). \tag{2.31}$$

See [Wic94].

Note the difference in (2.2.3.2) and (2.2.3.4) that self-duality no longer holds in (2.2.3.4) and the conditions are weakened.

$$\sum_k h(k) = \sqrt{2}; \tag{2.32}$$

$$\sum_k g(2k) = -\sum_k g(2k + 1); \tag{2.33}$$

$$\sum_k h'(k) = \sqrt{2}; \tag{2.34}$$

$$\sum_k g'(2k) = -\sum_k g'(2k + 1). \tag{2.35}$$

See [Wic94]. Having four operators gives plenty of freedom to construct filters with special properties, such as symmetry.

2.2.3.5 *Symlets*

The term of wavelet symlets are short for "symmetric wavelets." They are not perfectly symmetrical, but they are designed in such a way that they have the least asymmetry and highest number of vanishing moments for a given compact support [GW02]. Symlets are another family of Daubechies wavelets, thus are constructed in the same way as Daubechies wavelets.

Theorem 2.8. *The wavelet algorithms listed in this chapter can be realized as images; i.e., the wavelets as matrices can be applied to image matrices for wavelet decomposition.*

The proof is implemented in the next section.

2.3 Digital Image Representation and Mathematics Behind It

In this section, we will explore the digital image representation and the Mathematics behind it. MATLAB is an interactive system whose basic data element is an array that does not require dimensioning. This enables formulating solutions to many technical computing problems, especially those involving matrix representations, in a fraction of the time it would take to write a program in a scalar non-interactive language such as C or Fotran.

The name MATLAB stands for matrix laboratory. In university environments, MATLAB is the standard computational tool for introductory and advanced courses in mathematics, engineering, and science. In industry, MATLAB is the computational tool of choice for research, development, and analysis. MATLAB is complemented by a family of application-specific solutions called toolboxes; here, the Wavelet Toolbox is used [GWE04].

2.3.1 *Digital image representation*

An image is defined as a two-dimensional function i.e., a matrix, $f(x, y)$, where x and y are spatial coordinates, and the amplitude of f at any pair of coordinates (x, y) is called the intensity or gray level of the image at the point. Color images are formed by combining the individual two-dimensional images. For example, in the RGB color system, a color image consists of three, namely, red, green and blue, individual component images. Thus, many of the techniques developed for monochrome images can be extended to color images by processing the three component images individually. When x, y, and the amplitude values of f are all finite, discrete quantities, the image is called a digital image. The field of digital image processing refers to processing digital images by means of a digital computer. A digital image is composed of a finite number of elements, each of which has a particular location and value. These elements are referred to as picture elements, image elements, pels, and pixels. Since pixel is the most widely used term, the elements will be denoted as pixels from now on.

An image maybe continuous with respect to the x- and y-coordinates, and also in amplitude. Digitizing the coordinates as well as the amplitude will take into effect the conversion of such an image to digital form. Here, the digitization of the coordinate values are called sampling; digitizing the amplitude values is called quantization. A digital image is composed of a finite number of elements, each of which has a particular location and value. The field of digital image processing refers to processing digital images by means of a digital computer. See [GWE04].

2.3.1.1 *Coordinate convention*

Assume that an image $f(x, y)$ is sampled so that the resulting image has M rows and N columns. Then the image is of size $M \times N$. The values of the coordinates (x, y) are discrete quantities. Integer values are used for these discrete coordinates. In many image processing books, the image origin is set to be at $(x, y) = (0, 0)$. The next coordinate values along the first row of the image are $(x, y) = (0, 1)$. Note that the notation $(0, 1)$ is used to signify the second sample along the first row. These are not necessarily the actual values of physical coordinates when the image was sampled. Note that x ranges from 0 to $M - 1$, and y from 0 to $N - 1$, where x and y are integers. However, in the Wavelet Toolbox the notation (r, c) is used where r indicates rows and c indicates the columns. It could be noted that the order of coordinates is the same as the order discussed previously. Now, the major difference is that the origin of the coordinate system is at $(r, c) = (1, 1)$; hence, r ranges from 1 to M, and c from 1 to N for r and c integers. The coordinates are referred to as pixel coordinates. See [GWE04].

2.3.1.2 *Images as matrices*

The coordinate system discussed in the preceding section leads to the following representation for the digitized image function:

$$f(\mathbf{x}, \mathbf{y}) = \begin{bmatrix} f(0,0) & f(0,1) & \cdots & f(0, N-1) \\ f(1,0) & f(1,1) & \cdots & f(1, N-1) \\ \vdots & \vdots & \vdots & \vdots \\ f(M-1,0) & f(M-1,1) & \cdots & f(M-1, N-1) \end{bmatrix}.$$

The right-hand side of the equation is a representation of the digital image. Each element of this array (matrix) is called the pixel.

Now, in MATLAB, a digital image is represented as the following matrix:

$$
f = \begin{bmatrix}
f(1,1) & f(1,2) & \cdots & f(1,N) \\
f(2,1) & f(2,2) & \cdots & f(2,N) \\
\vdots & \vdots & \vdots & \vdots \\
f(M,1) & f(M,2) & \cdots & f(M,N)
\end{bmatrix},
\tag{2.36}
$$

where M = the number of rows and N = the number of columns. Matrices in MATLAB are stored in variables with names such as A, a, RGB, real array, and so on. See [GWE04].

2.3.1.3 *Color image representation in MATLAB*

An RGB color image is an $M \times N \times 3$ array or matrix of color pixels, where each color pixel consists of a triplet corresponding to the red, green, and blue components of an RGB image at a specific spatial location. An RGB image may be viewed as a "stack" of three gray-scale images, that when fed into the red, green, and blue inputs of a color monitor produce a color image on the screen. So from the "stack" of three images forming that RGB color image, each image is referred to as the red, green, and blue component images by convention. Now, the data class of the component images determine their range of values. If an RGB color image is of class double, meaning that all the pixel values are of type double, the range of values is $[0,1]$. Likewise, the range of values is $[0, 255]$ or $[0, 65535]$ for RGB images of class unit8 or unit16, respectively. The number of bits used to represent the pixel values of the component images determines the bit depth of an RGB color image. See [GWE04].

The RGB color space is shown graphically as an RGB color cube. The vertices of the cube are the primary (red, green, and blue) and secondary (cyan, magenta, and yellow) colors of light. See [GWE04].

2.3.1.4 *Indexed images*

An indexed image has two components: a data matrix of integers, X, and a colormap matrix, map. Matrix map is an $m \times 3$ array of class double containing floating-point values in the range $[0,1]$. The length, m, of the map is equal to the number of colors it defines. Each row of the map specifies the red, green, and blue components of a single color. An indexed image uses "direct mapping" of pixel intensity values of colormap values. The color of each pixel is determined by using the corresponding value of integer matrix X as a pointer into the map. If X is of class double, then all of its components with value 2 point to the second row, and so on. If X is of class unit 8 or unit 16, then all components with value 0 point to the first row in the map, all components with value 1 point to the second row, and so on [GWE04].

2.3.1.5 *The basics of color image processing*

Color image processing techniques deal with how the color images are handled for a variety of image-processing tasks. For the purposes of the following discussion, we subdivide color image processing into three principal areas: (1) color transformations (also called color mappings); (2) spatial processing of individual color planes; and (3) color vector processing. The first category deals with processing the pixels of each color plane based strictly on their values and not on their spatial coordinates. This category is analogous to the intensity transformations. The second category deals with spatial (neighborhood) filtering for individual color planes and is analogous to spatial filtering. The third category deals with techniques base on processing all components of a color image simultaneously. Since full-color images have at least three components, color pixels are indeed vectors. For example, in the RGB color images, the RGB system color point can be interpreted as a vector extending from the origin to that point in the RGB coordinate system.

Let c represent an arbitrary vector in RGB color space:

$$\mathbf{c} = \begin{bmatrix} c_R \\ c_G \\ c_B \end{bmatrix} = \begin{bmatrix} R \\ G \\ B \end{bmatrix}.$$

This equation indicates that the components of c are simply the RGB components of a color image at a point since the color components are a function of coordinates (x, y) by using the notation

$$\mathbf{c}(x, y) = \begin{bmatrix} c_R(x, y) \\ c_G(x, y) \\ c_B(x, y) \end{bmatrix} = \begin{bmatrix} R(x, y) \\ G(x, y) \\ B(x, y) \end{bmatrix}.$$

For an image of size $M \times N$, there are MN such vectors, $\mathbf{c}(\mathbf{x}, \mathbf{y})$, for $x = 0, 1, \ldots, M - 1$ and $y = 0, 1, \ldots, N - 1$ [GWE04].

In order for independent color component and vector-based processing to be equivalent, two conditions have to be satisfied: (i) the process has to be applicable to both vectors and scalars. (ii) the operation on each component of a vector must be independent of the other components. The averaging would be accomplished by summing the gray levels of all the pixels in the neighborhood. Or the averaging could be done by summing all the vectors in the neighborhood and dividing each component of the average vector as the sum of the pixels in the image corresponding to that component, which is the same as the result that would be obtained if the averaging were done on the neighborhood of each component image individually, and then the color vector were formed [GWE04].

2.3.2 *Reading images*

In MATLAB, images are read into the MATLAB environment using a function called imread. The syntax is as follows: imread(filename) Here, filename is a string containing the complete name of the image file including any applicable extension. For example, the command line $\gg f =$ imread (Barbara.png); reads the PNG image Barbara into image array or image matrix f.

Since there are three color components in the image, namely red, green, and blue components, the image is broken down into the three distinct color matrices $\mathbf{f}_R$, $\mathbf{f}_G$, and $\mathbf{f}_B$ in the form 2.36. See [GWE04].

2.3.3 *Wavelet decomposition of an image*

2.3.3.1 *Color conversion*

In the process of image compression, applying the compression to the RGB components of the image would result in undesirable color changes. Thus, the image is transformed into its intensity, hue, and color saturation components. The color transformation used in JPEG 2000 standard [SCE01] has been adopted. For the lossy compression, Equations (2.37) and (2.38) were used in the program.

$$\begin{bmatrix} Y \\ C_b \\ C_r \end{bmatrix} = \begin{bmatrix} 0.299 & 0.587 & 0.114 \\ -0.16875 & -0.33126 & 0.5 \\ 0.5 & -0.41869 & -0.08131 \end{bmatrix} \begin{bmatrix} R \\ G \\ B \end{bmatrix} \tag{2.37}$$

$$\begin{bmatrix} R \\ G \\ B \end{bmatrix} = \begin{bmatrix} 1.0 & 0 & 1.402 \\ 1.0 & 0.34413 & -0.71414 \\ 1.0 & 1.772 & 0 \end{bmatrix} \begin{bmatrix} Y \\ C_b \\ C_r \end{bmatrix}. \tag{2.38}$$

In YC_bC_r color space, Y is the single component that represents luminance. C_b and C_r store the color information where C_b stands for difference between the blue component and a reference value, and C_r is the difference between the red component and a reference value [GWE04]. In the case of lossless compression, Equations (2.39) and (2.40) were used.

$$\begin{bmatrix} Y_r \\ V_r \\ U_r \end{bmatrix} \begin{bmatrix} \lfloor \frac{R+2G+B}{4} \rfloor \\ R - G \\ B - G \end{bmatrix}. \tag{2.39}$$

$$\begin{bmatrix} G \\ R \\ B_r \end{bmatrix} = \begin{bmatrix} Y_r - \lfloor \frac{U_r+V_r}{4} \rfloor \\ V_r + G \\ U_r + G \end{bmatrix}. \tag{2.40}$$

Here, Y is the luminance and U and V are chrominance values (light intensity and color intensity), the subscript r stands for reversible. The advantage of this color system is that the human perception for the Y component is substantially more sensitive than for fluctuations in the U or V components. This can practically be used to transform U, and V components, which are transferred less. That is, of these components, it reduces the data set of these two components to $1/4$ of the original amount to be worth transferring [SCE01].

A 1-level wavelet transform of an $N \times M$ image can be represented as

$$\mathbf{f} \mapsto \begin{pmatrix} \mathbf{a}^1 & | & \mathbf{h}^1 \\ -- & & -- \\ \mathbf{v}^1 & | & \mathbf{d}^1 \end{pmatrix} \tag{2.41}$$

$$\mathbf{a}^1 = V_m^1 \otimes V_n^1 = \phi(x,y) = \phi(x)\phi(y) = \sum_i \sum_j h_i h_j \phi(2x-i)\phi(2y-j)$$
$$\mathbf{h}^1 = V_m^1 \otimes W_n^1 = \psi^H(x,y) = \psi(x)\phi(y) = \sum_i \sum_j g_i h_j \psi(2x-i)\phi(2y-j)$$
$$\mathbf{v}^1 = W_m^1 \otimes V_n^1 = \psi^V(x,y) = \phi(x)\psi(y) = \sum_i \sum_j h_i g_j \phi(2x-i)\psi(2y-j)$$
$$\mathbf{d}^1 = W_m^1 \otimes W_n^1 = \psi^D(x,y) = \psi(x)\psi(y) = \sum_i \sum_j g_i g_j \psi(2x-i)\psi(2y-j)$$
$$\tag{2.42}$$

where the subimages $\mathbf{h}^1, \mathbf{d}^1, \mathbf{a}^1$, and $\mathbf{v}^1$ each have the dimension of $N/2$ by $M/2$.

Here, $\mathbf{a}^1$ denotes the first averaged image, which consists of average intensity values of the original image. $\mathbf{h}^1$ denotes the first detailed image of horizontal components, which consists of intensity difference along the vertical axis of the original image. $\mathbf{v}^1$ denotes the first detailed image of vertical components, which consists of intensity difference along the horizontal axis of the original image. $\mathbf{d}^1$ denotes the first detailed image of diagonal components, which consists of intensity difference along the diagonal axis of the original image. The original image is reconstructed from the decomposed image by taking the sum of the averaged image and the detail images and scaling by a scaling factor. See [Wal99a].

Here, wavelet decomposition of images was performed the number of times the image size can be divided by 2, i.e., (floor(log2(min(size of Image)))) times. The averaged image of the previous level is decomposed into the four subimages in each level of wavelet image decomposition. This is related to the dyadic algorithm of MRA of wavelets where every thing is divided by 2 to decompose into the next level.

Applying further wavelet decomposition on image in Figure 2.3 would result in the next image in Figure 2.4. Note that the image on the top leftmost corner, namely average detail, gets blurrier as it gets "averaged" out and also note the horizontal, vertical, and diagonal components of the image. Further note the horizontal, vertical, and diagonal components explicitly shown in the horizontal detail, vertical detail, and diagonal detail. The horizontal, vertical, and diagonal components in the rectangular duster and the frame in the picture can be noted explicitly.

The following example illustrates a simple example where the average, horizontal, diagonal, and vertical components are explicitly depicted. As one can see, only the horizontal difference and some horizontalness are detected for the horizontal component and only the vertical difference and some verticalness are detected for the vertical component. As for the diagonal component, one can only see the diagonal difference and the average component carries the "shape" of the original image throughout.

The process of the computation using MATLAB keeps track of two, matrices C and S; C is the coefficient decomposition vector:

$$C = \begin{bmatrix} a(n) & h(n) & v(n) & d(n) & h(n-1) & \cdots & v(1) & d(1) \end{bmatrix},$$

where a, h, v and d are columnwise vectors containing approximation, horizontal, vertical, and diagonal coefficient matrices, respectively. C has $3n + 1$ sections where n is the number of wavelet decompositions. S is an

$(n + 2) \times 2$ bookkeeping matrix:

$$S = \begin{bmatrix} sa(n,:) & sd(n,:) & sd(n-1,:) & ... & sd(1,:) & sx \end{bmatrix},$$

where sa is the approximation size entry and sd is detail size entry [GWE04].

The above process is performed mathematically as follows: f_R, f_G, and f_B are treated as vectors of row vectors. For example, for fR we have

$$\boldsymbol{f}_R = \begin{bmatrix} f_{R_1} \\ f_{R_2} \\ \vdots \\ f_{R_M} \end{bmatrix} \quad \text{where } f_i = (f_{i,1}, f_{i,2}, \ldots, f_{i,N}). \qquad (2.43)$$

Then each row vector in f_R goes through the following operation:

$$\begin{bmatrix} s_{i,1} \\ s_{i,2} \\ \vdots \\ s_{i,N/2} \\ d_{i,1} \\ d_{i,2} \\ \vdots \\ d_{i,N/2} \end{bmatrix} = \begin{bmatrix} h_0 & h_1 & h_2 & h_3 & 0 & \cdots & \cdots & \cdots & 0 \\ 0 & 0 & h_0 & h_1 & h_2 & h_3 & 0 & \cdots & 0 \\ \cdots & \cdots & \cdots & \cdots & \cdots & \cdots & \cdots & \cdots & \cdots \\ g_0 & g_1 & g_2 & g_3 & 0 & \cdots & \cdots & \cdots & 0 \\ 0 & 0 & g_0 & g_1 & g_2 & g_3 & 0 & \cdots & 0 \\ \cdots & \cdots & \cdots & \cdots & \cdots & \cdots & \cdots & \cdots & \cdots \end{bmatrix} \begin{bmatrix} f_{i,1} \\ f_{i,2} \\ \vdots \\ f_{i,N} \end{bmatrix}.$$

$$(2.44)$$

After performing the above operation on each row vector, form a matrix with the resulting row vectors. Multiplying the same matrix to the column vector of the resulting matrix would result in (2.41). It can be seen in [PTVF92] that the above process is done by performing double loops of multiplication of the vector sequence with the matrix.

For the image compression, the quantization process takes place after the wavelet decomposition stage. That is, thresholding (Section 2.2.1.2) of the matrix takes place thus resulting in data reduction. A more detailed description of this process can be found in [Wal99a]. Also, see [BWP03] and [WB03]. Figure 2.9 describes the wavelet image compression. The wavelet decomposition step described above is the forward wavelet transform. Then thresholding takes place, which is another form of quantization,

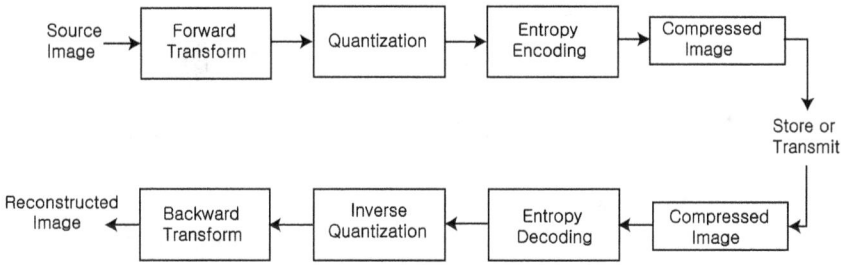

Fig. 2.9 Outline of the wavelet image compression process [SCE01].

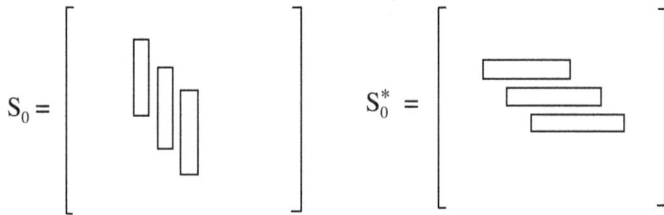

Fig. 2.10 S_0 and S_0^* as slanted infinite by infinite matrices.

followed by entropy encoding. Entropy encoding is further discussed in Chapter 7. Then to reconstruct back the image, entropy decoding is done, inverse quantization and backward wavelet transform are carried out resulting in reconstructed compressed image. Entropy encoding is discussed in Chapter 7 of this book.

2.3.4 *Mathematical insights*

In this section, we are going to make some changes. H and G in previous sections and chapters will now be denoted as S_0 and S_1. The projection operators S_0 and S_1 are operators in $L^2(\mathbb{R})$, since computers cannot compute infinite integrals we have to convert the mathematical decomposition process to something that can be expressed as a computer algorithm. This is made possible by the property that in Hilbert space $L^2(\mathbb{R}) \cong L^2(\mathbb{T}) \cong l^2(\mathbb{Z})$. See [Dou98]. Recall that choices are involved in the isomorphisms, e.g., $\cong L^2(\mathbb{T}) \cong l^2(\mathbb{Z})$ is Fourier's choice. Thus, an integral can be expressed in terms of sequences. The relationship between projection operators S_0 and S_1 can be expressed as the following matrices where the boxes are the coefficients h_j's as in (2.44).

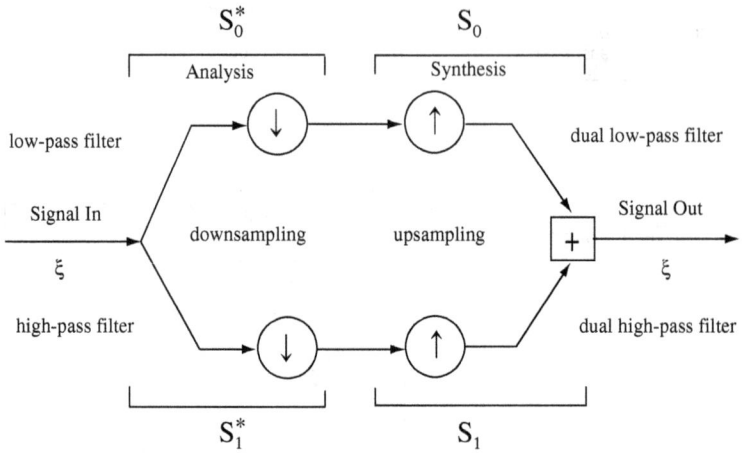

Fig. 2.11 For $L^2(\mathbb{T})$. Black box diagram of two frequency bands, with signal-in and signal-out.

V_n and W_n, defined in 2.2.2.1, can now be expressed as follows:

$$V_n = S_0 e_n$$

and

$$W_n = S_1 e_n,$$

where $e_n = (0, \ldots, 0, 1, 0, \ldots, 0)$, 1 is in the nth place.

We can compare the operations performed in two different spaces namely, $L^2(\mathbb{T})$ and $l^2(\mathbb{Z})$. See [Jor05] for more details about the operator notation.

Perfect reconstruction (operator notation):

$$S_0 S_0^* + S_1 S_1^* = I.$$

See (2.31), (2.44), and Figs. 2.11 and 2.12.

2.4 Haar Wavelet Matrix Example for Wavelet Decomposition and Reconstruction

This section presents an example of Haar wavelet decomposition of a matrix which illustrates wavelet digital image decomposition and reconstruction of a black and white image.

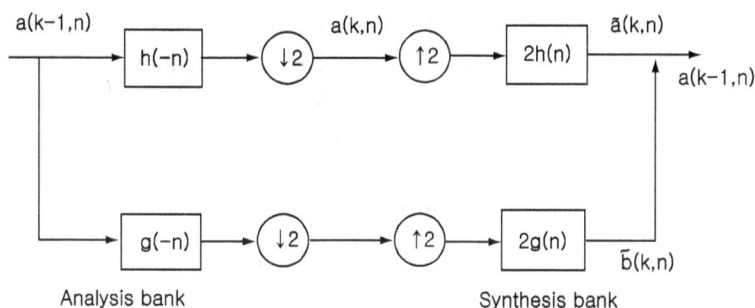

Fig. 2.12 For $l^2(\mathbb{Z})$.

The four Haar wavelet functions are defined on the real line $(-\infty, \infty)$ by

$$\varphi(t) = \begin{cases} 1 & 0 \le t < 1, \\ 0 & \text{otherwise.} \end{cases} \qquad \psi(t) = \begin{cases} 1 & 0 \le t < \frac{1}{2}, \\ -1 & \frac{1}{2} \le t < 1, \\ 0 & \text{otherwise.} \end{cases}$$

$$\psi_1(t) = \begin{cases} 1 & 0 \le t < \frac{1}{4}, \\ -1 & \frac{1}{4} \le t < \frac{1}{2}, \\ 0 & \text{otherwise.} \end{cases} \qquad \psi_2(t) = \begin{cases} -1 & \frac{1}{2} \le t < 1, \\ 0 & \text{otherwise.} \end{cases}$$

The function φ is known as the father wavelet, ψ, the mother wavelet, and ψ_1 and ψ_2 the daughter wavelets. We can create a vector representation of each of these functions for easier computation. In order to do this, since the functions are only supported on $0 \le t < 1$, we can create vectors in $\mathbb{R}^4$ where each entry represents the value of the function on $\frac{1}{4}$th of the interval. So, we have the first entry as the value of the function on $0 \le t < \frac{1}{4}$, the second entry as the value of the function on $\frac{1}{4} \le t < \frac{1}{2}$, the third entry as the value of the function on $\frac{1}{2} \le t < \frac{3}{4}$, and the fourth entry as the value of the function on $\frac{3}{4} \le t < 1$.

Example 2.9. This leads us to the following set of vectors in $\mathbb{R}^4$, please see [AS99]:

$$\begin{bmatrix} 1 \\ 1 \\ 1 \\ 1 \end{bmatrix}, \begin{bmatrix} 1 \\ 1 \\ -1 \\ -1 \end{bmatrix}, \begin{bmatrix} 1 \\ -1 \\ 0 \\ 0 \end{bmatrix}, \begin{bmatrix} 0 \\ 0 \\ -1 \\ -1 \end{bmatrix}.$$

We can easily see that the Haar wavelet vectors form an orthogonal set by looking at their dot products, which should equal zero. To check orthogonality:

$$(1,1,1,1) \cdot (1,1,-1,-1) = 1 + 1 + (-1) + (-1) = 0,$$

$$(1,1,1,1) \cdot (1,-1,0,0) = 1 + (-1) + 0 + 0 = 0,$$

$$(1,1,1,1) \cdot (0,0,1,-1) = 0 + 0 + 1 + (-1) = 0,$$

$$(1,1,-1,-1) \cdot (1,-1,0,0) = 1 + (-1) + 0 + 0 = 0,$$

$$(1,1,-1,-1) \cdot (0,0,1,-1) = 0 + 0 + 1 + (-1) = 0,$$

$$(1,-1,0,0) \cdot (0,0,1,-1) = 0 + 0 + 0 + 0 = 0.$$

Now we see it as an orthonorm system. If we want an orthonormal system, we need to normalize each of these vectors by dividing each of them by their corresponding norm, as follows:

$$\frac{(1,1,1,1)}{\sqrt{1^2 + 1^2 + (1)^2 + (1)^2}} = \frac{(1,1,1,1)}{2} = \left(\frac{1}{2}, \frac{1}{2}, \frac{1}{2}, \frac{1}{2}\right),$$

$$\frac{(1,1,-1,-1)}{\sqrt{1^2 + 1^2 + (-1)^2 + (-1)^2}} = \frac{(1,1,-1,-1)}{2} = \left(\frac{1}{2}, \frac{1}{2}, -\frac{1}{2}, -\frac{1}{2}\right),$$

$$\frac{(1,-1,0,0)}{\sqrt{1^2 + (-1)^2 + 0^2 + 0^2}} = \frac{(1,-1,0,0)}{\sqrt{2}} = \left(\frac{1}{\sqrt{2}}, -\frac{1}{\sqrt{2}}, 0, 0\right),$$

$$\frac{(0,0,1,-1)}{\sqrt{0^2 + 0^2 + 1^2 + (-1)^2}} = \frac{(0,0,1,-1)}{\sqrt{2}} = \left(0, 0, \frac{1}{\sqrt{2}}, -\frac{1}{\sqrt{2}}\right).$$

Now we can see that we have an orthonormal set of vectors by checking their dot products, which should equal 1:

$$\left(\frac{1}{2}, \frac{1}{2}, \frac{1}{2}, \frac{1}{2}\right) \cdot \left(\frac{1}{2}, \frac{1}{2}, \frac{1}{2}, \frac{1}{2}\right) = \frac{1}{4} + \frac{1}{4} + \frac{1}{4} + \frac{1}{4} = 1,$$

$$\left(\frac{1}{2}, \frac{1}{2}, -\frac{1}{2}, -\frac{1}{2}\right) \cdot \left(\frac{1}{2}, \frac{1}{2}, -\frac{1}{2}, -\frac{1}{2}\right) = \frac{1}{4} + \frac{1}{4} + \frac{1}{4} + \frac{1}{4} = 1,$$

$$\left(\frac{1}{\sqrt{2}}, -\frac{1}{\sqrt{2}}, 0, 0\right) \cdot \left(\frac{1}{\sqrt{2}}, -\frac{1}{\sqrt{2}}, 0, 0\right) = \frac{1}{2} + \frac{1}{2} + 0 + 0 = 1,$$

$$\left(0, 0, \frac{1}{\sqrt{2}}, -\frac{1}{\sqrt{2}}\right) \cdot \left(0, 0, \frac{1}{\sqrt{2}}, -\frac{1}{\sqrt{2}}\right) = 0 + 0 + \frac{1}{2} + \frac{1}{2} = 1.$$

Now that we have an orthonormal basis, we can take a one-dimensional vector of length four, and write it as a linear product of the orthonormal vectors. Take the vector $(2, 10, 7, 1)$, for example:

$$\left(\frac{1}{2}, \frac{1}{2}, \frac{1}{2}, \frac{1}{2}\right) \cdot (2, 10, 7, 1) = 1 + 5 + \frac{7}{2} + \frac{1}{2} = 10,$$

$$\left(\frac{1}{2}, \frac{1}{2}, -\frac{1}{2}, -\frac{1}{2}\right) \cdot (2, 10, 7, 1) = 1 + 5 + \left(-\frac{7}{2}\right) + \left(-\frac{1}{2}\right) = 2,$$

$$\left(\frac{1}{\sqrt{2}}, -\frac{1}{\sqrt{2}}, 0, 0\right) \cdot (2, 10, 7, 1) = \frac{2}{\sqrt{2}} + \left(-\frac{10}{\sqrt{2}}\right) + 0 + 0$$
$$= \sqrt{2} - 5\sqrt{2} = -4\sqrt{2},$$

$$\left(0, 0, \frac{1}{\sqrt{2}}, -\frac{1}{\sqrt{2}}\right) \cdot (2, 10, 7, 1) = 0 + 0 + \frac{7}{\sqrt{2}} + \left(-\frac{1}{\sqrt{2}}\right)$$
$$= \frac{7}{2}\sqrt{2} - \frac{1}{2}\sqrt{2} = 3\sqrt{2}.$$

We can take our results and create a new vector $\left(10, 2, -4\sqrt{2}, 3\sqrt{2}\right)$. We can now use this new vector to reconstruct our original vector by using our orthonormal basis:

$$10 \begin{bmatrix} \frac{1}{2} \\ \frac{1}{2} \\ \frac{1}{2} \\ \frac{1}{2} \end{bmatrix} + 2 \begin{bmatrix} \frac{1}{2} \\ \frac{1}{2} \\ -\frac{1}{2} \\ -\frac{1}{2} \end{bmatrix} + (-4\sqrt{2}) \begin{bmatrix} \frac{1}{\sqrt{2}} \\ -\frac{1}{\sqrt{2}} \\ 0 \\ 0 \end{bmatrix} + 3\sqrt{2} \begin{bmatrix} 0 \\ 0 \\ \frac{1}{\sqrt{2}} \\ -\frac{1}{\sqrt{2}} \end{bmatrix}$$

$$= \begin{bmatrix} 5 \\ 5 \\ 5 \\ 5 \end{bmatrix} + \begin{bmatrix} 1 \\ 1 \\ -1 \\ -1 \end{bmatrix} + \begin{bmatrix} -4 \\ 4 \\ 0 \\ 0 \end{bmatrix} + \begin{bmatrix} 0 \\ 0 \\ 3 \\ -3 \end{bmatrix} = \begin{bmatrix} 2 \\ 10 \\ 7 \\ 1 \end{bmatrix}.$$

Note that we can also decompose our original vector using a two component basis to get the same new vector. Take the basis $\left(\frac{1}{\sqrt{2}}, \frac{1}{\sqrt{2}}\right), \left(\frac{1}{\sqrt{2}}, -\frac{1}{\sqrt{2}}\right)$. We then decompose as follows:

$$\left(\frac{1}{\sqrt{2}}, \frac{1}{\sqrt{2}}\right) \cdot (2, 10) = \frac{2}{\sqrt{2}} + \frac{10}{\sqrt{2}} = \frac{12}{\sqrt{2}} = 6\sqrt{2},$$

$$\left(\frac{1}{\sqrt{2}}, \frac{1}{\sqrt{2}}\right) \cdot (7, 1) = \frac{7}{\sqrt{2}} + \frac{1}{\sqrt{2}} = \frac{8}{\sqrt{2}} = 4\sqrt{2}.$$

Now create a new vector $\left(6\sqrt{2}, 4\sqrt{2}\right)$ and do the following:

$$\left(\frac{1}{\sqrt{2}}, \frac{1}{\sqrt{2}}\right) \cdot \left(6\sqrt{2}, 4\sqrt{2}\right) = 6 + 4 = 10,$$

$$\left(\frac{1}{\sqrt{2}}, -\frac{1}{\sqrt{2}}\right) \cdot \left(6\sqrt{2}, 4\sqrt{2}\right) = 6 - 4 = 2.$$

Now we have our first two entries in our new vector. To get our last two entries in our new matrix do the following:

$$\left(\frac{1}{\sqrt{2}}, -\frac{1}{\sqrt{2}}\right) \cdot (2, 10) = \frac{2}{\sqrt{2}} - \frac{10}{\sqrt{2}} = -\frac{8}{\sqrt{2}} = -4\sqrt{2},$$

$$\left(\frac{1}{\sqrt{2}}, -\frac{1}{\sqrt{2}}\right) \cdot (7, 1) = \frac{7}{\sqrt{2}} - \frac{1}{\sqrt{2}} = \frac{6}{\sqrt{2}} = 3\sqrt{2}.$$

Now we have our last two entries. Putting our results together, we see we have the same new vector $\left(10, 2, -4\sqrt{2}, 3\sqrt{2}\right)$ which we have previously reconstructed.

2.4.1 *Examples*

In order to understand the wavelet decomposition and reconstruction, let's look at a one-dimensional vector of length 2^n and apply the H and G operators where H is the average operator and G is the difference operator. H and G operators are matrices where their size is dependent on the size of the vector on which H and G are operating. If the vector is $1 \times m$, then the size of both H and G will be $m \times \frac{m}{2}$. When applying H and G to a vector, it will obviously create new vectors which are half of the length of the original vector. If we are going to apply H and G to a vector of length 8, then

$$H = \begin{bmatrix} \frac{1}{\sqrt{2}} & \frac{1}{\sqrt{2}} & 0 & 0 & 0 & 0 & 0 & 0 \\ 0 & 0 & \frac{1}{\sqrt{2}} & \frac{1}{\sqrt{2}} & 0 & 0 & 0 & 0 \\ 0 & 0 & 0 & 0 & \frac{1}{\sqrt{2}} & \frac{1}{\sqrt{2}} & 0 & 0 \\ 0 & 0 & 0 & 0 & 0 & 0 & \frac{1}{\sqrt{2}} & \frac{1}{\sqrt{2}} \end{bmatrix}$$

and

$$G = \begin{bmatrix} \frac{1}{\sqrt{2}} & -\frac{1}{\sqrt{2}} & 0 & 0 & 0 & 0 & 0 & 0 \\ 0 & 0 & \frac{1}{\sqrt{2}} & -\frac{1}{\sqrt{2}} & 0 & 0 & 0 & 0 \\ 0 & 0 & 0 & 0 & \frac{1}{\sqrt{2}} & -\frac{1}{\sqrt{2}} & 0 & 0 \\ 0 & 0 & 0 & 0 & 0 & 0 & \frac{1}{\sqrt{2}} & -\frac{1}{\sqrt{2}} \end{bmatrix}.$$

We follow the patterns to create H and G to accommodate longer vectors. We can see that applying H to a signal $\mathbf{a} = (a_1, a_2, a_3, a_4, a_5, a_6, a_7, a_8, \ldots)$ will give us the vector $H\mathbf{a} = \left(\frac{a_1+a_2}{\sqrt{2}}, \frac{a_3+a_4}{\sqrt{2}}, \frac{a_5+a_6}{\sqrt{2}}, \frac{a_7+a_8}{\sqrt{2}}, \ldots \right)$ and applying G to that same signal will give us $G\mathbf{a} = \left(\frac{a_1-a_2}{\sqrt{2}}, \frac{a_3-a_4}{\sqrt{2}}, \frac{a_5-a_6}{\sqrt{2}}, \frac{a_7-a_8}{\sqrt{2}}, \ldots \right)$.

Consider $f = (5, 3, -2, 8, 0, 8, -3, 7)$.

$$Hf = \left(\frac{5+3}{\sqrt{2}}, \frac{-2+8}{\sqrt{2}}, \frac{0+8}{\sqrt{2}}, \frac{-3+7}{\sqrt{2}} \right) = (4\sqrt{2}, 3\sqrt{2}, 4\sqrt{2}, 2\sqrt{2})$$

and

$$Gf = \left(\frac{5-3}{\sqrt{2}}, \frac{-2-8}{\sqrt{2}}, \frac{0-8}{\sqrt{2}}, \frac{-3-7}{\sqrt{2}} \right) = \left(\sqrt{2}, -5\sqrt{2}, -4\sqrt{2}, -5\sqrt{2} \right).$$

Let us call our new vectors $\mathbf{c}^1$ and $\mathbf{d}^1$, respectively. If we apply H and G again to $\mathbf{c}^1$, we get

$$H\mathbf{c}^1 = \left(\frac{4\sqrt{2}+3\sqrt{2}}{\sqrt{2}}, \frac{4\sqrt{2}+2\sqrt{2}}{\sqrt{2}} \right) = (7, 6) \quad \text{and}$$

$$G\mathbf{c}^1 = \left(\frac{4\sqrt{2}-3\sqrt{2}}{\sqrt{2}}, \frac{4\sqrt{2}-2\sqrt{2}}{\sqrt{2}} \right) = (1, 2).$$

Call these vectors $\mathbf{c}^2$ and $\mathbf{d}^2$, respectively. Applying the operators one more time will give us

$$H\mathbf{c}^2 = \left(\frac{7+6}{\sqrt{2}} \right) = \left(\frac{13}{\sqrt{2}} \right) = \left(\frac{13\sqrt{2}}{2} \right) \quad \text{and}$$

$$G\mathbf{c}^2 = \left(\frac{7-6}{\sqrt{2}} \right) = \left(\frac{1}{\sqrt{2}} \right) = \left(\frac{\sqrt{2}}{2} \right).$$

Call these $\mathbf{c}^3$ and $\mathbf{d}^3$, respectively.

Now in order to get back to our original signal $\mathbf{f}$, we will need adjoint operators H^* and G^* which will create the identity $HH^* + GG^* = 1$. In order to get this identity, we define H^* and G^* as the transposes of H and G, respectively. Applying these to a signal $\mathbf{b} = (b_1, b_2, b_3, b_4, \ldots)$ will give us

$$\mathbf{b}H^* = \left(\frac{b_1}{\sqrt{2}}, \frac{b_1}{\sqrt{2}}, \frac{b_2}{\sqrt{2}}, \frac{b_2}{\sqrt{2}}, \frac{b_3}{\sqrt{2}}, \frac{b_3}{\sqrt{2}}, \frac{b_4}{\sqrt{2}}, \frac{b_4}{\sqrt{2}}, \cdots \right)$$

and

$$\mathbf{b}G^* = \left(\frac{b_1}{\sqrt{2}}, -\frac{b_1}{\sqrt{2}}, \frac{b_2}{\sqrt{2}}, -\frac{b_2}{\sqrt{2}}, \frac{b_3}{\sqrt{2}}, -\frac{b_3}{\sqrt{2}}, \frac{b_4}{\sqrt{2}}, -\frac{b_4}{\sqrt{2}}, \cdots \right).$$

To begin the reconstruction stage, apply H^* to $\mathbf{c}^3$ and G^* to $\mathbf{d}^3$.

$$\mathbf{c}^3 H^* = \left(\frac{13}{2}, \frac{13}{2} \right) \quad \text{and} \quad \mathbf{d}^3 G^* = \left(\frac{1}{2}, -\frac{1}{2} \right)$$

We can see that adding these together gets us

$$\mathbf{c}^3 H^* + \mathbf{c}^3 G^* = \left(\frac{13}{2}, \frac{13}{2} \right) + \left(\frac{1}{2}, -\frac{1}{2} \right) = (7, 6)$$

which is our $\mathbf{c}^2$ vector. We can continue this process until we get to our original signal:

$$\mathbf{c}^2 H^* = \left(\frac{7}{\sqrt{2}}, \frac{7}{\sqrt{2}}, \frac{6}{\sqrt{2}}, \frac{6}{\sqrt{2}} \right) \quad \text{and} \quad \mathbf{d}^2 G^* = \left(\frac{1}{\sqrt{2}}, -\frac{1}{\sqrt{2}}, \frac{2}{\sqrt{2}}, -\frac{2}{\sqrt{2}} \right)$$

then

$$\mathbf{c}^2 H^* + \mathbf{d}^2 G^* = \left(\frac{7}{\sqrt{2}}, \frac{7}{\sqrt{2}}, \frac{6}{\sqrt{2}}, \frac{6}{\sqrt{2}} \right) + \left(\frac{1}{\sqrt{2}}, -\frac{1}{\sqrt{2}}, \frac{2}{\sqrt{2}}, -\frac{2}{\sqrt{2}} \right)$$

$$= (4\sqrt{2}, 3\sqrt{2}, 4\sqrt{2}, 2\sqrt{2}) = \mathbf{c}^1$$

$$\mathbf{c}^1 H^* = (4, 4, 3, 3, 4, 4, 2, 2) \quad \text{and}$$

$$\mathbf{d}^1 G^* = (1, -1, -5, 5, -4, 4, -5, 5)$$

then

$$\mathbf{c}^1 H^* + \mathbf{d}^1 G^* = (4, 4, 3, 3, 4, 4, 2, 2) + (1, -1, -5, 5, -4, 4, -5, 5)$$

$$= (5, 3, -2, 8, 0, 8, -3, 7)$$

which is our original signal $\mathbf{f}$.

Now let us take a look at a two-dimensional example. This will be useful when applied to an image, since an image can be represented as a two-dimensional vector. In this case, we will be applying our average and difference operators to the rows and the columns so we must have an average and difference operator for both row multiplication and column multiplication.

Example 2.10. Let

$$
H_r = \begin{bmatrix}
\frac{1}{\sqrt{2}} & \frac{1}{\sqrt{2}} & 0 & 0 & 0 & 0 & 0 & \cdots \\
0 & 0 & \frac{1}{\sqrt{2}} & \frac{1}{\sqrt{2}} & 0 & 0 & 0 & \cdots \\
0 & 0 & 0 & 0 & \frac{1}{\sqrt{2}} & \frac{1}{\sqrt{2}} & 0 & \cdots \\
\vdots & \vdots & \vdots & \vdots & \vdots & \vdots & \vdots & \vdots
\end{bmatrix}
$$

$$
H_c = \begin{bmatrix}
\frac{1}{\sqrt{2}} & 0 & 0 & \cdots \\
\frac{1}{\sqrt{2}} & 0 & 0 & \cdots \\
0 & \frac{1}{\sqrt{2}} & 0 & \cdots \\
0 & \frac{1}{\sqrt{2}} & 0 & \cdots \\
0 & 0 & \frac{1}{\sqrt{2}} & \cdots \\
0 & 0 & \frac{1}{\sqrt{2}} & \cdots \\
0 & 0 & 0 & \frac{1}{\sqrt{2}} \\
\vdots & \vdots & \vdots & \vdots
\end{bmatrix}
$$

$$
G_r = \begin{bmatrix}
\frac{1}{\sqrt{2}} & -\frac{1}{\sqrt{2}} & 0 & 0 & 0 & 0 & 0 & \cdots \\
0 & 0 & \frac{1}{\sqrt{2}} & -\frac{1}{\sqrt{2}} & 0 & 0 & 0 & \cdots \\
0 & 0 & 0 & 0 & \frac{1}{\sqrt{2}} & -\frac{1}{\sqrt{2}} & 0 & \cdots \\
\vdots & \vdots & \vdots & \vdots & \vdots & \vdots & \vdots & \vdots
\end{bmatrix}
$$

$$
G_c = \begin{bmatrix}
\frac{1}{\sqrt{2}} & 0 & 0 & \cdots \\
-\frac{1}{\sqrt{2}} & 0 & 0 & \cdots \\
0 & \frac{1}{\sqrt{2}} & 0 & \cdots \\
0 & -\frac{1}{\sqrt{2}} & 0 & \cdots \\
0 & 0 & \frac{1}{\sqrt{2}} & \cdots \\
0 & 0 & -\frac{1}{\sqrt{2}} & \cdots \\
0 & 0 & 0 & \frac{1}{\sqrt{2}} \\
\vdots & \vdots & \vdots & \vdots
\end{bmatrix}
$$

Given a matrix

$$
F = \begin{bmatrix}
f_{1,1} & f_{2,1} & f_{3,1} & \cdots \\
f_{1,2} & f_{2,2} & f_{3,2} & \cdots \\
f_{1,3} & f_{2,3} & f_{3,3} & \cdots \\
\vdots & \vdots & \vdots & \vdots
\end{bmatrix},
$$

we have $H_r F H_c$, the average, $H_r F G_c$, the horizontal difference, $G_r F H_c$, the vertical difference, and $G_r F G_c$, the diagonal difference. We can then construct a new matrix $\begin{bmatrix} H_r F H_c & H_r F G_c \\ G_r F H_c & G_r F G_c \end{bmatrix}$. Let us take the following matrix M and decompose it as described.

$$
M = \begin{bmatrix}
3 & 4 & -1 & 0 \\
-5 & 10 & 7 & -3 \\
1 & 0 & 8 & 6 \\
0 & 0 & 2 & 1
\end{bmatrix}
$$

$$
H_r M H_c = \begin{bmatrix}
\frac{1}{\sqrt{2}} & \frac{1}{\sqrt{2}} & 0 & 0 \\
0 & 0 & \frac{1}{\sqrt{2}} & \frac{1}{\sqrt{2}}
\end{bmatrix}
\begin{bmatrix}
3 & 4 & -1 & 0 \\
-5 & 10 & 7 & -3 \\
1 & 0 & 8 & 6 \\
0 & 0 & 2 & 1
\end{bmatrix}
$$

$$
\begin{bmatrix}
\frac{1}{\sqrt{2}} & 0 \\
\frac{1}{\sqrt{2}} & 0 \\
0 & \frac{1}{\sqrt{2}} \\
0 & \frac{1}{\sqrt{2}}
\end{bmatrix}
= \begin{bmatrix}
6 & \frac{3}{2} \\
\frac{1}{2} & \frac{17}{2}
\end{bmatrix} = A
$$

$$
H_r M G_c = \begin{bmatrix}
\frac{1}{\sqrt{2}} & \frac{1}{\sqrt{2}} & 0 & 0 \\
0 & 0 & \frac{1}{\sqrt{2}} & \frac{1}{\sqrt{2}}
\end{bmatrix}
\begin{bmatrix}
3 & 4 & -1 & 0 \\
-5 & 10 & 7 & -3 \\
1 & 0 & 8 & 6 \\
0 & 0 & 2 & 1
\end{bmatrix}
$$

$$
\begin{bmatrix}
\frac{1}{\sqrt{2}} & 0 \\
-\frac{1}{\sqrt{2}} & 0 \\
0 & \frac{1}{\sqrt{2}} \\
0 & -\frac{1}{\sqrt{2}}
\end{bmatrix}
= \begin{bmatrix}
-8 & \frac{9}{2} \\
\frac{1}{2} & \frac{3}{2}
\end{bmatrix} = H
$$

$$G_r M H_c = \begin{bmatrix} \frac{1}{\sqrt{2}} & -\frac{1}{\sqrt{2}} & 0 & 0 \\ 0 & 0 & \frac{1}{\sqrt{2}} & -\frac{1}{\sqrt{2}} \end{bmatrix} \begin{bmatrix} 3 & 4 & -1 & 0 \\ -5 & 10 & 7 & -3 \\ 1 & 0 & 8 & 6 \\ 0 & 0 & 2 & 1 \end{bmatrix}$$

$$\begin{bmatrix} \frac{1}{\sqrt{2}} & 0 \\ \frac{1}{\sqrt{2}} & 0 \\ 0 & \frac{1}{\sqrt{2}} \\ 0 & \frac{1}{\sqrt{2}} \end{bmatrix} = \begin{bmatrix} 1 & -\frac{5}{2} \\ \frac{1}{2} & \frac{11}{2} \end{bmatrix} = V$$

$$G_r M G_c = \begin{bmatrix} \frac{1}{\sqrt{2}} & -\frac{1}{\sqrt{2}} & 0 & 0 \\ 0 & 0 & \frac{1}{\sqrt{2}} & -\frac{1}{\sqrt{2}} \end{bmatrix} \begin{bmatrix} 3 & 4 & -1 & 0 \\ -5 & 10 & 7 & -3 \\ 1 & 0 & 8 & 6 \\ 0 & 0 & 2 & 1 \end{bmatrix}$$

$$\begin{bmatrix} \frac{1}{\sqrt{2}} & 0 \\ -\frac{1}{\sqrt{2}} & 0 \\ 0 & \frac{1}{\sqrt{2}} \\ 0 & -\frac{1}{\sqrt{2}} \end{bmatrix} = \begin{bmatrix} 7 & -\frac{11}{2} \\ \frac{1}{2} & \frac{1}{2} \end{bmatrix} = D$$

and now we can construct our new matrix

$$\begin{bmatrix} 6 & \frac{3}{2} & -8 & \frac{9}{2} \\ \frac{1}{2} & \frac{17}{2} & \frac{1}{2} & \frac{3}{2} \\ 1 & -\frac{5}{2} & 7 & -\frac{11}{2} \\ \frac{1}{2} & \frac{11}{2} & \frac{1}{2} & \frac{1}{2} \end{bmatrix}.$$

Now to reconstruct our original matrix, we compute $H_r^* A H_c^*, H_r^* HG_c^*, G_r^* V H_c^*$, and $G_r^* D G_c^*$, then add each of those together and we should get our original matrix:

$$H_r^* A H_c^* = \begin{bmatrix} \frac{1}{\sqrt{2}} & 0 \\ \frac{1}{\sqrt{2}} & 0 \\ 0 & \frac{1}{\sqrt{2}} \\ 0 & \frac{1}{\sqrt{2}} \end{bmatrix} \begin{bmatrix} 6 & \frac{3}{2} \\ \frac{1}{2} & \frac{17}{2} \end{bmatrix}$$

$$\begin{bmatrix} \frac{1}{\sqrt{2}} & \frac{1}{\sqrt{2}} & 0 & 0 \\ 0 & 0 & \frac{1}{\sqrt{2}} & \frac{1}{\sqrt{2}} \end{bmatrix} = \begin{bmatrix} 3 & 3 & \frac{3}{4} & \frac{3}{4} \\ 3 & 3 & \frac{3}{4} & \frac{3}{4} \\ \frac{1}{4} & \frac{1}{4} & \frac{17}{4} & \frac{17}{4} \\ \frac{1}{4} & \frac{1}{4} & \frac{17}{4} & \frac{17}{4} \end{bmatrix},$$

$$H_r^* H G_c^* = \begin{bmatrix} \frac{1}{\sqrt{2}} & 0 \\ \frac{1}{\sqrt{2}} & 0 \\ 0 & \frac{1}{\sqrt{2}} \\ 0 & \frac{1}{\sqrt{2}} \end{bmatrix} \begin{bmatrix} -8 & \frac{9}{2} \\ \frac{1}{2} & \frac{3}{2} \end{bmatrix}$$

$$\begin{bmatrix} \frac{1}{\sqrt{2}} & -\frac{1}{\sqrt{2}} & 0 & 0 \\ 0 & 0 & \frac{1}{\sqrt{2}} & -\frac{1}{\sqrt{2}} \end{bmatrix} = \begin{bmatrix} -4 & 4 & \frac{9}{4} & -\frac{9}{4} \\ -4 & 4 & \frac{9}{4} & -\frac{9}{4} \\ \frac{1}{4} & -\frac{1}{4} & \frac{3}{4} & -\frac{3}{4} \\ \frac{1}{4} & -\frac{1}{4} & \frac{3}{4} & -\frac{3}{4} \end{bmatrix},$$

$$G_r^* V H_c^* = \begin{bmatrix} \frac{1}{\sqrt{2}} & 0 \\ -\frac{1}{\sqrt{2}} & 0 \\ 0 & \frac{1}{\sqrt{2}} \\ 0 & -\frac{1}{\sqrt{2}} \end{bmatrix} \begin{bmatrix} 1 & -\frac{5}{2} \\ \frac{1}{2} & \frac{11}{2} \end{bmatrix}$$

$$\begin{bmatrix} \frac{1}{\sqrt{2}} & \frac{1}{\sqrt{2}} & 0 & 0 \\ 0 & 0 & \frac{1}{\sqrt{2}} & \frac{1}{\sqrt{2}} \end{bmatrix} = \begin{bmatrix} \frac{1}{2} & \frac{1}{2} & -\frac{5}{4} & -\frac{5}{4} \\ -\frac{1}{2} & -\frac{1}{2} & \frac{5}{4} & \frac{5}{4} \\ \frac{1}{4} & \frac{1}{4} & \frac{11}{4} & \frac{11}{4} \\ -\frac{1}{4} & -\frac{1}{4} & -\frac{11}{4} & -\frac{11}{4} \end{bmatrix},$$

$$G_r^* D G_c^* = \begin{bmatrix} \frac{1}{\sqrt{2}} & 0 \\ -\frac{1}{\sqrt{2}} & 0 \\ 0 & \frac{1}{\sqrt{2}} \\ 0 & -\frac{1}{\sqrt{2}} \end{bmatrix} \begin{bmatrix} 7 & -\frac{11}{2} \\ \frac{1}{2} & \frac{1}{2} \end{bmatrix}$$

$$\begin{bmatrix} \frac{1}{\sqrt{2}} & -\frac{1}{\sqrt{2}} & 0 & 0 \\ 0 & 0 & \frac{1}{\sqrt{2}} & -\frac{1}{\sqrt{2}} \end{bmatrix} = \begin{bmatrix} \frac{7}{2} & -\frac{7}{2} & -\frac{11}{4} & \frac{11}{4} \\ -\frac{7}{2} & \frac{7}{2} & \frac{11}{4} & -\frac{11}{4} \\ \frac{1}{4} & -\frac{1}{4} & \frac{1}{4} & -\frac{1}{4} \\ -\frac{1}{4} & \frac{1}{4} & -\frac{1}{4} & \frac{1}{4} \end{bmatrix}.$$

$$H_r^* A H_c^* + H_r^* H G_c^* + G_r^* V H_c^* + G_r^* D G_c^*$$

$$
= \begin{bmatrix} 3 & 3 & \frac{3}{4} & \frac{3}{4} \\ 3 & 3 & \frac{3}{4} & \frac{3}{4} \\ \frac{1}{4} & \frac{1}{4} & \frac{17}{4} & \frac{17}{4} \\ \frac{1}{4} & \frac{1}{4} & \frac{17}{4} & \frac{17}{4} \end{bmatrix} + \begin{bmatrix} -4 & 4 & \frac{9}{4} & -\frac{9}{4} \\ -4 & 4 & \frac{9}{4} & -\frac{9}{4} \\ \frac{1}{4} & -\frac{1}{4} & \frac{3}{4} & -\frac{3}{4} \\ \frac{1}{4} & -\frac{1}{4} & \frac{3}{4} & -\frac{3}{4} \end{bmatrix}
$$

$$
+ \begin{bmatrix} \frac{1}{2} & \frac{1}{2} & -\frac{5}{4} & -\frac{5}{4} \\ -\frac{1}{2} & -\frac{1}{2} & \frac{5}{4} & \frac{5}{4} \\ \frac{1}{4} & \frac{1}{4} & \frac{11}{4} & \frac{11}{4} \\ -\frac{1}{4} & -\frac{1}{4} & -\frac{11}{4} & -\frac{11}{4} \end{bmatrix} + \begin{bmatrix} \frac{7}{2} & -\frac{7}{2} & -\frac{11}{4} & \frac{11}{4} \\ -\frac{7}{2} & \frac{7}{2} & \frac{11}{4} & -\frac{11}{4} \\ \frac{1}{4} & -\frac{1}{4} & \frac{1}{4} & -\frac{1}{4} \\ -\frac{1}{4} & \frac{1}{4} & -\frac{1}{4} & \frac{1}{4} \end{bmatrix}
$$

$$
= \begin{bmatrix} 3 & 4 & -1 & 0 \\ -5 & 10 & 7 & -3 \\ 1 & 0 & 8 & 6 \\ 0 & 0 & 2 & 1 \end{bmatrix},
$$

which is our original matrix M.

2.5 Wavelet Color Image Compression Results and Discussion

In this section, we discuss the wavelet color image compression results and make comparisons based on wavelets discussed in Section 2.2.3 using different threshold values which have been defined in Section 2.2.1.2.

2.5.1 *Implementation of the program*

The program was implemented using MATLAB with various subroutines that enable the wavelet transformation, image compression, and threshold computation from the Wavelet Toolkit.

2.5.2 *Discussion of the results*

2.5.2.1 *Wavelet lossy image compression*

There are various factors that influence image compression. As mentioned above in Section 2.2, nonorthogonality of the wavelet may cause the compression to be lossy. When threshold is applied to the compression, some of the "insignificant" coefficients are thrown out, thus resulting in lossy compression. Also, the number of levels the wavelet transform is applied to would influence the compression quality. Although the lossiness caused by the nonorthogonal wavelet was not avoidable when certain wavelets were used, an attempt to minimize the lossiness was made for the number of levels part by going down all the way to the single-pixel level when the wavelet transform was applied (floor(log2(min(size of Image)))). In addition, various threshold values are applied to observe the lossiness.

A lossy compression method tends to produce inaccuracies in the decompressed image. Lossy compression method is used when these inaccuracies are so small that they are imperceptible. If those imperceptible inaccuracies are acceptable, the lossy technique is advantageous compared to the lossless ones for higher compression ratios can be attained.

In order to support the claims made by comparison of the resulting images and the theoretical knowledge that we obtained from the texts, some common numerical comparisons for images are made. They are the compression ratio, the root mean square error, rms, the relative two norm difference, D, and the peak signal-to-noise ratio, PSNR. The formulas used are as follows:

$$\text{Compression ratio} = \frac{1}{\frac{X \times Y \times 3 - (L^2 \text{norm recovery in } \% X \times Y \times 3/100)}{X \times Y \times 3}} \tag{2.45}$$

$$\text{rms} = \sqrt{\frac{\sum_{n=1}^{3} \sum_{i=1}^{Y} \sum_{j=1}^{X} (f_{i,j,n} - g_{i,j,n})^2}{X \times Y \times 3}} \tag{2.46}$$

$$D = \sqrt{\frac{\sum_{n=1}^{3} \sum_{i=1}^{Y} \sum_{j=1}^{X} (f_{i,j,n} - g_{i,j,n})^2}{\sum_{n=1}^{3} \sum_{i=1}^{Y} \sum_{j=1}^{X} f_{i,j,n}^2}} \tag{2.47}$$

$$\text{PSNR} = 20 - \log \frac{255}{\text{rms}}. \tag{2.48}$$

See [Wal99a].

The peak signal-to-noise ratio, PSNR, measures the quality of the compressed image in comparison to the original image. What it measures is the peak error. 255 is the maximum pixel value of an 8-bit image. When comparing image compression, PSNR is an approximation to human perception of reconstruction quality. So the higher the PSNR, the higher the quality of the compression image.

Mean square error is the cumulative squared error between the compressed and the original image. The lower the value of mean square error, the smaller the error. The relative two norm difference, D, which is self-explanatory from the formula, measures the norm difference of sum of difference squares of each pixel to the compressed and original image. So the smaller the relative two-norm difference is, closer the images are to each other.

2.5.2.2 *Barbara image compression using different wavelets and threshold values*

A selected number of various wavelet transforms: Coiflets (Coif1, Coif3, and Coif5), symlets (Sym2, Sym5, and Sym8), Daubechies wavelets (db1, db2, and db3) and biorthogonal wavelets (bior1.1, bior2.2, and bior5.5) with three different threshold values (5, 10, and 20) were used to compress the 8-bit color image Barbara.png (Section 2.3.3) to illustrate the performance results of image compression depending on wavelet types and threshold values. The image compression results are presented as follows:

A square-sized image is chosen as a test image because it gives optimal number of wavelet decompositions as mentioned in Section 2.3.3, the number of decomposition level is determined by dividing the size of the image matrix by 2. Since an image is 2-D, having the same number of rows and columns gives optimal number of levels of wavelet decomposition. This leads to maximal number of subdivisions of average, horizontal, vertical, and diagonal details. This results from the dyadic algorithm of MRA of wavelets where a vector is divided by 2 to decompose into the next level. This is illustrated as a simple matrix example in Section 2.4, Chapter 3, and Section 2.3.3 in theory.

Fig. 2.13 The original Barbara image from the test image pool.

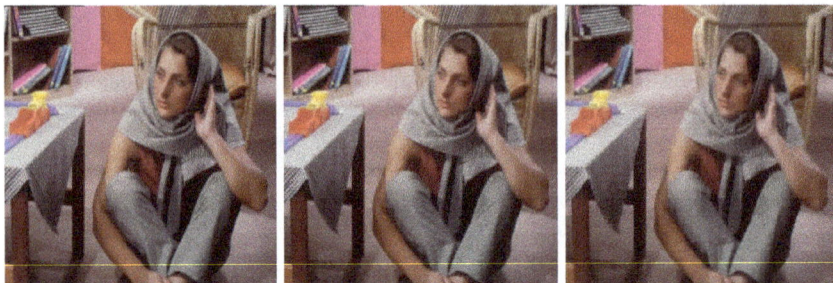

Fig. 2.14 From left to right: Haar wavelet, db1 Barbara image compression with threshold values 5, 10, and 20.

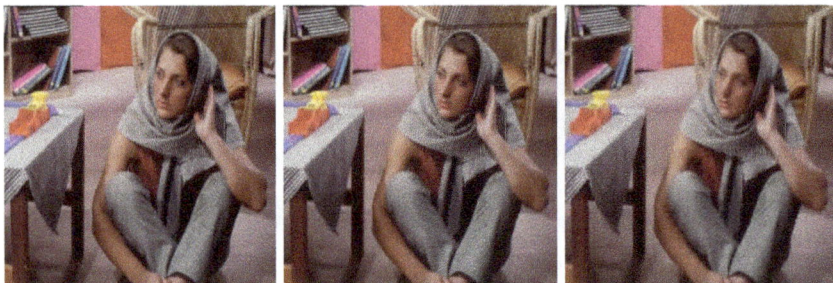

Fig. 2.15 From left to right: Daubechies wavelet, db2 Barbara image compression with threshold values 5, 10, and 20.

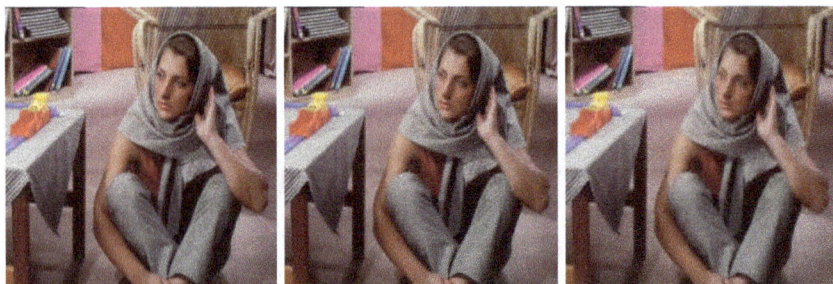

Fig. 2.16 From left to right: Daubechies wavelet, db3 Barbara image compression with threshold values 5, 10, and 20.

Fig. 2.17 From left to right: Biorthogonal wavelet, Bior1.1 Barbara image compression with threshold values 5, 10, and 20.

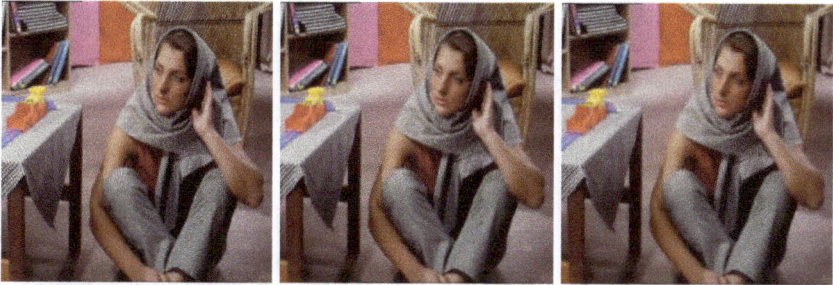

Fig. 2.18 From left to right: Biorthogonal wavelet, Bior2.2 Barbara image compression with threshold values 5, 10, and 20.

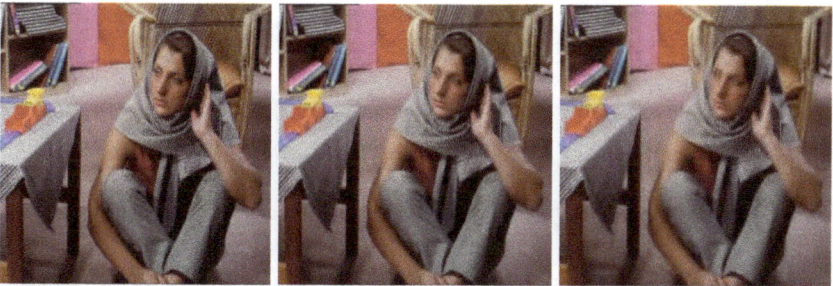

Fig. 2.19 From left to right: Biorthogonal wavelet, Bior5.5 Barbara image compression with threshold values 5, 10, and 20.

Based on observations of compressed images, it is noted that images compressed with smaller threshold value 5, in comparison to 10 and 20, appear closer to the original Barbara.png. The compressed images using

Fig. 2.20 From left to right: Coiflet, Coif1 Barbara image compression with threshold values 5, 10, and 20.

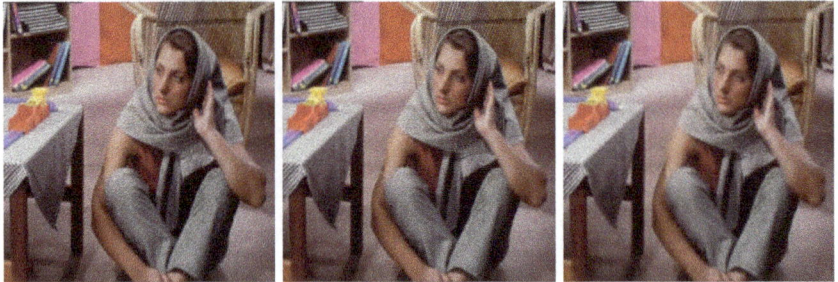

Fig. 2.21 From left to right: Coiflet, Coif3 Barbara image compression with threshold values 5, 10, and 20.

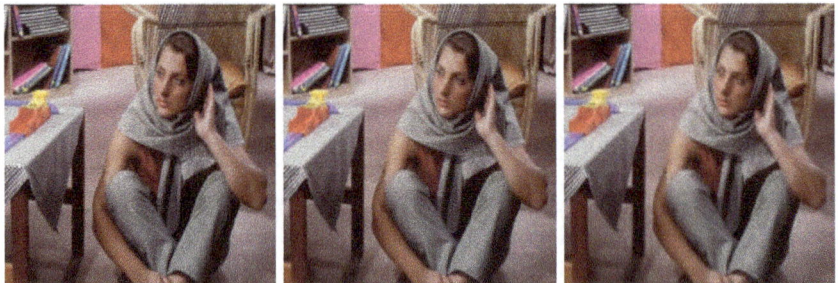

Fig. 2.22 From left to right: Coiflet, Coif5 Barbara image compression with threshold values 5, 10, and 20.

threshold value 20 look blurrier than the other images compressed using threshold values 5 and 10.

Also, the comparison tables above show that the smaller the threshold value, the smaller the root mean square error (rms), the peak signal-to-noise

Fig. 2.23 From left to right: Symlet, Sym2 Barbara image compression with threshold values 5, 10, and 20.

Fig. 2.24 From left to right: Symlet, Sym5 Barbara image compression with threshold values 5, 10, and 20.

Fig. 2.25 From left to right: Symlet, Sym8 Barbara image compression with threshold values 5, 10, and 20.

ratio, PSNR, is higher, and the relative two-norm difference, D is lower. This indicates that the compressed image has less error, higher quality, and is closer to the original image. It can also be observed that the compression ratio is smaller as less pixels are thresholded.

Table 2.1.   Table for Barbara wavelet compression using threshold value 5.

Wavelet	RMS	D	PSNR	Ratio
Coif1	4.0547	0.0332	35.9716	32.0975
Coif3	3.7887	0.0311	36.5611	21.6768
Coif5	3.7501	0.0307	36.6499	13.7314
bior1.1	4.4901	0.0368	35.0856	25.1965
bior2.2	4.0417	0.0331	35.9994	31.8597
bior5.5	4.3355	0.0355	35.3901	34.8249
Sym2	4.0742	0.0334	35.9300	33.7824
Sym5	3.8341	0.0314	36.4576	29.6730
Sym8	3.7722	0.0309	36.5989	23.4065
db1	4.4901	0.0368	35.0856	25.1965
db2	4.0742	0.0334	35.9300	33.7824
db3	3.9514	0.0324	36.1957	33.8286

Table 2.2.   Table for Barbara wavelet compression using threshold value 10.

Wavelet	RMS	D	PSNR	Ratio
Coif1	6.4328	0.0527	31.9629	58.2349
Coif3	5.9699	0.0489	32.6114	33.3402
Coif5	5.8983	0.0483	32.7162	19.7733
bior1.1	7.1552	0.0586	31.0384	52.9477
bior2.2	6.2471	0.0512	32.2172	57.9981
bior5.5	6.8867	0.0564	31.3705	59.3464
Sym2	6.4644	0.0530	31.9202	63.3540
Sym5	6.0436	0.0495	32.5048	50.7689
Sym8	5.9453	0.0487	32.6474	37.2218
db1	7.1552	0.0586	31.0384	52.9477
db2	6.4644	0.0530	31.9202	63.3540
db3	6.2305	0.0511	32.2404	60.7005

Now, looking at the performances of each of the wavelet transforms given the same threshold value, bior2.2 (Biorthogonal wavelet), sym5 (Symlet), and Coif3 (Coiflet) seem to have produced the less flawless compressed images compared to all the other wavelets.

Within the Daubechies wavelets, db2 and db3 appear to have produced the least flawless compressed image; that agrees with what was discussed above in Daubechies wavelets that db2 is better in signal compression than db1(Haar). See Section 2.3.3. Considering the error values, compression

Table 2.3. Table for Barbara wavelet compression using threshold value 20.

Wavelet	RMS	D	PSNR	Ratio
Coif1	9.7625	0.0800	28.3396	108.2137
Coif3	9.0822	0.0744	28.9670	52.1844
Coif5	8.9902	0.0737	29.0554	29.8045
bior1.1	10.8370	0.0888	27.4326	110.8900
bior2.2	9.1202	0.0748	28.9307	103.1862
bior5.5	10.5322	0.0863	27.6805	102.8941
Sym2	9.8040	0.0804	28.3028	116.2732
Sym5	9.2065	0.0755	28.8489	86.5173
Sym8	9.0504	0.0742	28.9974	59.5653
db1	10.8370	0.0888	27.4326	110.8900
db2	9.8040	0.0804	28.3028	116.2732
db3	9.4551	0.0775	28.6175	107.4413

ratios and the compressed image quality, sym5 appears to be the best choice for the image compression among the various wavelets used. Also, having the extra properties as mentioned under the Coiflets section, Section 2.2.3.3, Coif3 performed generally better in image compression compared to other wavelets. Having the biorthogonal property also seems to result in better image compression. On the other hand, the orthogonal Daubechies wavelets do not seem to perform better than coiflets, biorthogonal wavelets, and symlets. See [Wal99a].

Also, having longer support, which is proportional to the order of the wavelet, appears to worsen the performance of the image compression.

Wavelet compression did show remarkable performance, especially with smaller threshold value; it was not differentiable in between the original image and the compressed image for some cases. The average, horizontal, vertical, and diagonal details at multiple levels from MRS in wavelet image decomposition play a major role in image compression. In Section 5.9.2, we present image compression using principal component analysis of a black and white image, just to illustrate how image compression is differently done using the idea of principal component with correlation of pixel values.

Chapter 3

Wavelets as Multiresolutions

In Chapter 2, we discussed discrete wavelet transforms explicitly and their applications in wavelet image compression. In this chapter we outline a number of direct links between the two cases of wavelet analysis, continuous and discrete. The theme of the first is perhaps best known, for example the creation of compactly supported wavelets in $L^2(\mathbb{R}^n)$ with suitable properties such as localization, vanishing moments, and differentiability. The second (discrete) deals with computation, with sparse matrices, and with algorithms for encoding digitized information such as speech and images. This is centered on constructive approaches to subdivision-filters, to their matrix representation (by sparse matrices), and corresponding fast algorithms. For both approaches, we outline computational transforms; but our emphasis is on effective and direct links between computational analysis of discrete filters, on the one hand, and on continuous wavelets, on the other. By the latter we include both $L^2(\mathbb{R}^n)$-analysis as well as fractal analysis. To facilitate the discussion of the interplay between discrete (used by engineers) and continuous (harmonic analysis), we include a list of terminology commonly used in the two areas; and we include comments on translation between them. The novice might find the following references helpful: [Coh03a, JMR01, Son06b].

3.1 Multiresolutions: History, Applications, Examples, Discussion, and Algorithms

Haar's work in 1909–1910 had implicitly the key idea which got wavelet mathematics started on a roll 75 years later with Yves Meyer, Ingrid

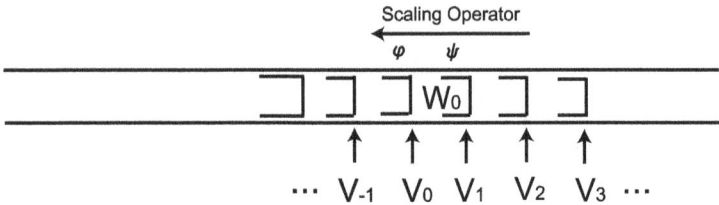

Fig. 3.1 Multiresolution. $L^2(\mathbb{R}^d)$-version (continuous); $\varphi \in V_0$, $\psi \in W_0$.

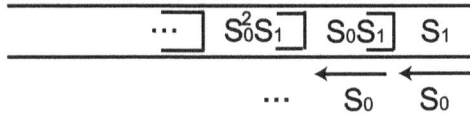

Fig. 3.2 Multiresolution. $l^2(\mathbb{Z})$-version (discrete); $\varphi \in V_0$, $\psi \in W_0$. Graphic representation of scales of resolution and intermediate detail spaces. The nested boxes in Figure 3.1 are represented by isometries and projections in this figure. The analogous representation of resolution scales for image compression is treated in detail in Section 4.2.

Daubechies, Stéphane Mallat, and others [CDDD03, Coh03b, Dau92, GM85, JMR01, Law99, Mar82] — namely the idea of a multiresolution. In that respect Haar was ahead of his time. See Figures 3.1 and 3.2 for details.

$$\cdots \subset V_{-1} \subset V_0 \subset V_1 \subset \cdots, \quad V_0 + W_0 = V_1$$

The word "multiresolution" suggests a connection to optics from physics. So that should have been a hint to mathematicians to take a closer look at trends in signal and image processing! Moreover, even staying within mathematics, it turns out that as a general notion this same idea of a "multiresolution" has long roots in mathematics, even in such modern and pure areas as operator theory and Hilbert-space geometry. Looking even closer at these interconnections, we can now recognize scales of subspaces (so-called multiresolutions) in classical algorithmic construction of orthogonal bases in inner-product spaces, now taught in lots of mathematics courses under the name of the Gram–Schmidt algorithm. Indeed, a closer look at good old Gram–Schmidt reveals that it is a matrix algorithm, hence, a new mathematical tool involving non-commutativity!

If the signal to be analyzed is an image, then why not select a fixed but suitable *resolution* (or a subspace of signals corresponding to a selected resolution), and then do the computations there? The selection of a fixed "resolution" is dictated by practical concerns. That idea was key in turning computation of wavelet coefficients into iterated matrix algorithms. As the matrix operations get large, the computation is carried out in a variety of paths arising from big matrix products. The dichotomy, continuous vs. discrete, is quite familiar to engineers. The industrial engineers typically work with huge volumes of numbers.

Numbers! — So why wavelets? Well, what matters to the industrial engineer is not really the wavelets, but the fact that special wavelet functions serve as an efficient way to encode large data sets — I mean encode for computations. And the wavelet algorithms are computational. They work on numbers. Encoding numbers into pictures, images, or graphs of functions comes later, perhaps at the very end of the computation. But without the graphics, I doubt that we would understand any of this half as well as we do now. The same can be said for the many issues that relate to the crucial mathematical concept of self-similarity, as we know it from fractals, and more generally from recursive algorithms.

3.2 Glossary

This *Glossary* consists of a list of terms used: In mathematics, in probability, in engineering, and on occasion in physics. To clarify the seemingly confusing use of up to four different names for the same idea or concept, we have further added informal explanations spelling out the reasons behind the differences in current terminology from neighboring fields. Also, see [Dau92, Bri10a, Bri10b, Bri13a, Bri13b, Sha93a, Sha57, dLMSS56, HKMMT89, GM85, JMR01, Mey00a, Coh03a, CDDD03, Coh03b, CCM16, AMWW16, CD04, DDLY00, Don00] for more details.

Disclaimer: This glossary has the structure of four columns. A number of terms are listed line by line, and each line is followed by explanation. Some "terms" have up to four separate (yet commonly accepted) names.

Mathematics	Probability	Engineering	Physics
Function (Measurable)	Random Variable	Signal	State

Mathematically, functions may map between any two sets, say, from X to Y; but if X is a probability space (typically called Ω), it comes with a σ-algebra $\mathcal{B}$ of measurable sets, and probability measure P. Elements E in $\mathcal{B}$ are called events, and $P(E)$, the probability of E. Corresponding measurable functions with values in a vector space are called random variables, a terminology which suggests a stochastic viewpoint. The function values of a random variable may represent the outcomes of an experiment, for example, "throwing of a die."

Yet, function theory is widely used in engineering where functions are typically thought of as signals. In this case, X may be the real line for time, or $\mathbb{R}^d$. Engineers visualize functions as signals. A particular signal may have a stochastic component, and this feature simply introduces an extra stochastic variable into the "signal," for example noise.

Turning to physics, in our present application, the physical functions will be typically in some L^2-space, and L^2-functions with unit norm represent quantum mechanical "states."

Sequence (Incl. Vector-Valued)	Random Walk	Time-Series	Measurement

Mathematically, a sequence is a function defined on the integers $\mathbb{Z}$ or on subsets of $\mathbb{Z}$, for example, the natural numbers $\mathbb{N}$. Hence, if time is discrete, this to the engineer represents a time series, such as a speech signal, or any measurement which depends on time. But we will also allow functions on lattices such as $\mathbb{Z}^d$.

In the case $d = 2$, we may be considering the grayscale numbers which represent exposure in a digital camera. In this case, the function (grayscale) is defined on a subset of $\mathbb{Z}^2$, and is then simply a matrix.

A random walk on $\mathbb{Z}^d$ is an assignment of a sequential and random motion as a function of time. The randomness presupposes assigned probabilities. But we will use the term "random walk" also in connection with random walks on combinatorial trees.

Nested Subspaces	Refinement	Multiresolution	Scales of Visual Resolutions

While finite or infinite families of nested subspaces are ubiquitous in mathematics, and have been popular in Hilbert space theory for generations (at least since the 1930s), this idea was revived in a different guise in 1986 by Stéphane Mallat, then an engineering graduate student. In its adaptation to wavelets, the idea is now referred to as the multiresolution method.

Readers are referred to Appendix A for basics on operators in Hilbert space and Appendix B for basics on wavelets and multiresolutions. What made the idea especially popular in the wavelet community was that it offered a skeleton on which various discrete algorithms in applied mathematics could be attached and turned into wavelet constructions in harmonic analysis. In fact, what we now call multiresolutions have come to signify a crucial link between the world of discrete wavelet algorithms, which are popular in computational mathematics and in engineering (signal/image processing, data mining, etc.), on the one hand, and on the other hand continuous wavelet bases in function spaces, especially in $L^2(\mathbb{R}^d)$. Further, the multiresolution idea closely mimics how fractals are analyzed with the use of finite function systems.

But in mathematics, or more precisely in operator theory, the underlying idea dates back to the work of John von Neumann, Norbert Wiener, and Herman Wold, where nested and closed subspaces in Hilbert space were used extensively in an axiomatic approach to stationary processes, especially for time series. Wold proved that any (stationary) time series can be decomposed into two different parts: The first (deterministic) part can be exactly described by a linear combination of its own past, while the second part is the opposite extreme; it is *unitary*, in the language of von Neumann.

von Neumann's version of the same theorem is a pillar in operator theory. It states that every isometry in a Hilbert space $\mathcal{H}$ is the unique sum of a shift isometry and a unitary operator, i.e., the initial Hilbert space $\mathcal{H}$ splits canonically as an orthogonal sum of two subspaces $\mathcal{H}_s$ and $\mathcal{H}_u$ in $\mathcal{H}$, one which carries the shift operator, and the other $\mathcal{H}_u$ the unitary part. The shift isometry is defined from a nested scale of closed spaces V_n, such that the intersection of these spaces is $\mathcal{H}_u$. Specifically,

$$\cdots \subset V_{-1} \subset V_0 \subset V_1 \subset V_2 \subset \cdots \subset V_n \subset V_{n+1} \subset \cdots$$

$$\bigwedge_n V_n = \mathcal{H}_u, \quad \text{and} \quad \bigvee_n V_n = \mathcal{H}.$$

However, Stéphane Mallat was motivated instead by the notion of scales of resolutions in the sense of optics. This in turn is based on a certain "artificial-intelligence" approach to vision and optics, developed earlier by David Marr at MIT, an approach which imitates the mechanism of vision in the human eye.

The connection from these developments in the 1980s back to von Neumann is this: Each of the closed subspaces V_n corresponds to a level of resolution in such a way that a larger subspace represents a finer resolution. Resolutions are relative, not absolute! In this view, the relative complement of the smaller (or coarser) subspace in larger space then represents the visual detail which is added in passing from a blurred image to a finer one, i.e., to a finer visual resolution.

This view became an instant hit in the wavelet community, as it offered a repository for the fundamental father and the mother functions, also called the scaling function φ and the wavelet function ψ, respectively. Via a system of translation and scaling operators, these functions then generate nested subspaces, and we recover the scaling identities which initialize the appropriate algorithms. What results is now called the family of pyramid algorithms in wavelet analysis. The approach itself is called the multiresolution approach (MRA) to wavelets. And in the meantime various generalizations (GMRAs) have emerged.

In all of this, there was a second "accident" at play: As it turned out, pyramid algorithms in wavelet analysis now lend themselves via multiresolutions, or nested scales of closed subspaces, to an analysis based on frequency bands. Here we refer to bands of frequencies as they have already been used for a long time in signal processing.

One reason for the success in varied disciplines of the same geometric idea is perhaps that it is closely modeled on how we historically have represented numbers in the positional number system. Analogies to the Euclidean algorithm seem especially compelling.

Operator	Process	Black Box	Observable (If Self-adjoint)

In linear algebra, students are familiar with the distinctions between (linear) transformations T (here called "operators") and matrices. For a fixed operator $T: V \to W$, there is a variety of matrices, one for each choice of basis in V and in W. In many engineering applications, the transformations are not restricted to be linear, but instead represent some experiment ("black box," in Norbert Wiener's terminology), one with an input and an output, usually functions of time. The input could be an external voltage function, the black box, an electric circuit, and the output, the resulting voltage in the circuit. (The output is a solution to a differential equation.)

This context is somewhat different from that of quantum mechanical (QM) operators $T: V \to V$ where V is a Hilbert space. In QM, self-adjoint operators represent observables such as position Q and momentum P, or time and energy.

Fourier Dual Pair	Generating Function	Time/Frequency P/Q

The following dual pairs position Q/momentum P, and time/energy may be computed with the use of Fourier series or Fourier transforms; and in this sense they are examples of Fourier dual pairs. If for example time is discrete, then frequency may be represented by numbers in the interval $[0, 2\pi)$; or in $[0, 1)$ if we enter the number 2π into the Fourier exponential. Functions of the frequency are then periodic, so the two endpoints are identified. In the case of the interval $[0, 1)$, 0 on the left is identified with 1 on the right. So a low frequency band is an interval centered at 0, while a high frequency band is an interval centered at 1/2. Let a function W on $[0, 1)$ represent a probability assignment. Such functions W are thought of as "filters" in signal processing. We say that W is low-pass if it is 1 at 0, or if it is near 1 for frequencies near 0. Low-pass filters pass signals with low frequencies, and block the others.

If instead some filter W is 1 at $1/2$, or takes values near 1 for frequencies near $1/2$, then we say that W is high-pass; it passes signals with high frequency.

Convolution	—	Filter	Smearing

Pointwise multiplication of functions of frequencies corresponds in the Fourier dual time-domain to the operation of convolution (or of Cauchy product if the time-scale is discrete). The process of modifying a signal with a fixed convolution is called a linear filter in signal processing. The corresponding Fourier dual frequency function is then referred to as "frequency response" or the "frequency response function."

More generally, in the continuous case, since convolution tends to improve smoothness of functions, physicists call it "smearing."

Decomposition (e.g., Fourier Coefficients in a expansion)	—	Analysis	Frequency Components

---

Calculating the Fourier coefficients is "analysis," and adding up the pure frequencies (i.e., summing the Fourier series) is called synthesis. But this view carries over more generally to engineering where there are more operations involved on the two sides, e.g., breaking up a signal into its frequency bands, transforming further, and then adding up the "banded" functions in the end. If the signal out is the same as the signal in, we say that the analysis/synthesis yields perfect reconstruction.

Integrate (e.g., Inverse Fourier Transform)	Reconstruct	Synthesis	Superposition

---

Here the terms related to "synthesis" refer to the second half of the kind of signal-processing design outlined in the previous paragraph.

Subspace	—	Resolution	(Signals in a) Frequency Band

For a space of functions (signals), the selection of certain frequencies serves as a way of selecting special signals. When the process of scaling is introduced into optics of a digital camera, we note that a nested family of subspaces corresponds to a grading of visual resolutions.

Cuntz Relations	—	Perfect Reconstruction from Subbands	Subband Decomposition

$$\sum_{i=0}^{N-1} S_i S_i^* = 1, \quad \text{and} \quad S_i^* S_j = \delta_{i,j} 1.$$

Inner Product	Correlation	Transition Probability from One State to Another	Probability of Transition

In many applications, a vector space with inner products captures perfectly the geometric and probabilistic features of the situation. This can be axiomatized in the language of Hilbert space; and the inner product is the most crucial ingredient in the familiar axiom system for Hilbert space.

$f_{\text{out}} = T f_{\text{in}}$	—	Input/Output	Transformation of States

Systems theory language for operators $T: V \to W$. Then vectors in V are inputs, and the range of T, the output.

Fractal	—	—	—

Intuitively, think of a fractal as reflecting similarity of scales such as is seen in fern-like images that look "roughly" the same at small and at large scales. Fractals are produced from an infinite iteration of a finite set of maps, and this algorithm is perfectly suited to the kind of subdivision which is a cornerstone of the discrete wavelet algorithm. Self-similarity could refer alternately to space, and to time. And further versatility is added, in that flexibility is allowed into the definition of "similar."

— — **Data Mining** —

The problem of how to handle and make use of large volumes of data is a corollary of the digital revolution. As a result, the subject of data mining itself changes rapidly. Digitized information (data) is now easy to capture automatically and to store electronically. In science, in commerce, and in industry, data represent collected observations and information: In business, there are data on markets, competitors, and customers. In manufacturing, there are data for optimizing production opportunities, and for improving processes. A tremendous potential for data mining exists in medicine, genetics, and energy. But raw data are not always directly usable, as is evident by inspection. A key to advances is our ability to *extract information and knowledge* from the data (hence "data mining"), and to understand the phenomena governing data sources. Data mining is now taught in a variety of forms in engineering departments, as well as in statistics and computer science departments.

One of the structures often hidden in data sets is some degree of *scale*. The goal is to detect and identify one or more natural global and local scales in the data. Once this is done, it is often possible to detect associated similarities of scale, much like the familiar scale-similarity from multidimensional wavelets, and from fractals. Indeed, various adaptations of wavelet-like algorithms have been shown to be useful. These algorithms themselves are useful in *detecting* scale-similarities, and are applicable to other types of pattern recognition. Hence, in this context, generalized multiresolutions offer another tool for discovering structures in large data sets, such as those stored in the resources of the Internet. Because of the sheer volume of data involved, a strictly manual analysis is out of the question. Instead, sophisticated query processors based on statistical and mathematical techniques are used in generating insights and extracting conclusions from data sets.

Chapter 4

Discrete and Continuous Wavelet Transforms

Continuing from Chapter 3, in this chapter we outline several points of view on the interplay between discrete and continuous wavelet transforms; stressing both pure and applied aspects. We outline some new links between the two transform technologies based on the theory of representations of generators and relations. By this we mean a finite system of generators which are represented by operators in Hilbert space. We further outline how these representations yield subband filter banks for signal and image processing algorithms. This chapter is inspired by [JS14a]. This builds more mathematical theory to discrete wavelet transforms discussed in Chapter 3 with wavelet image compression examples.

The word "wavelet transform" (WT) means different things to different people: Pure and applied mathematicians typically give different answers to the question "What is the WT?" And engineers in turn have their own preferred quite different approach to WTs. Still there are two main trends in how WTs are used, the *continuous* WT on one side, and the *discrete* WT on the other. Here, we offer a user-friendly outline of both, but with a slant toward geometric methods from the theory of operators in Hilbert space. The novice might find the following references helpful: [CDDD03, DJ06b, DJ06c].

In this chapter, we begin with the glossary (Section 3.2). This is a substantial part of our account, and it reflects the multiplicity of how the subject is used.

The concept of multiresolutions or multiresolution analysis (MRA) serves as a link between the discrete and continuous theory.

In Section 4.8, we summarize how different mathematicians and scientists have contributed to and shaped the subject over the years.

The next two sections then offer a technical overview of both the discrete and the continuous WTs. This includes basic tools from Fourier analysis and from operators in Hilbert space. In Sections 4.5 and 4.6, we outline the connections between the separate parts of mathematics and their applications to WTs.

Readers are referred to Appendix A for basics on operators in Hilbert space and Appendix B for basics on wavelets and multiresolutions.

4.1 Introduction

While applied problems such as time series, signals, and processing of digital images come from engineering and from the sciences, they have in the past two decades taken a life of their own as an exciting new area of applied mathematics. While searches in Google on these keywords typically yield sites numbered in the millions, the diversity of applications is wide, and it seems reasonable here to narrow our focus to some of the approaches that are both more mathematical and more recent. For references, see, e.g., [AK06b, BLM06, Liu06, SN96]. In addition, our own interests (e.g., [Jor03, Jor06a, Son06a, Son06b]) have colored the following presentation. Each of the two areas, the discrete side and the continuous theory, is huge as measured by recent journal publications. A leading theme in our chapter is the independent interest in a multitude of interconnections between the discrete algorithms and their uses in the more mathematical analysis of function spaces (continuous wavelet transforms). The mathematics involved in the study and the applications of this interaction we feel is of benefit to both mathematicians and to engineers. See also [Jor03]. An early paper [DL92] by Daubechies and Lagarias was especially influential in connecting the two worlds, discrete and continuous.

4.2 The Discrete vs. Continuous Wavelet Algorithms

4.2.1 *The discrete wavelet transform*

If one stays with function spaces, it is then popular to pick the d-dimensional Lebesgue measure on $\mathbb{R}^d$, $d = 1, 2, \ldots$, and pass to the Hilbert space $L^2(\mathbb{R}^d)$ of all square integrable functions on $\mathbb{R}^d$, referring to the d-dimensional Lebesgue measure. A wavelet basis refers to a family of basis functions for

$L^2(\mathbb{R}^d)$ generated from a finite set of normalized functions ψ_i, the index i chosen from a fixed and finite index set I, and from two operations, one called scaling, and the other translation. The scaling is typically specified by a d by d matrix over the integers $\mathbb{Z}$ such that all the eigenvalues in modulus are bigger than one, lie outside the closed unit disk in the complex plane. The d-lattice is denoted $\mathbb{Z}^d$, and the translations will be by vectors selected from $\mathbb{Z}^d$. We say that we have a wavelet basis if the triple indexed family $\psi_{i,j,k}(x) := |\det A|^{j/2}\psi(A^j x + k)$ forms an orthonormal basis (ONB) for $L^2(\mathbb{R}^d)$ as i varies in I, $j \in \mathbb{Z}$, and $k \in \mathbb{R}^d$. The word "orthonormal" for a family F of vectors in a Hilbert space $\mathcal{H}$ refers to the norm and the inner product in $\mathcal{H}$: The vectors in an orthonormal family F are assumed to have norm one, and to be mutually orthogonal. If the family is also total (i.e., the vectors in F span a subspace which is dense in $\mathcal{H}$), we say that F is an orthonormal basis (ONB) [JS14a].

While there are other popular wavelet bases, e.g., frame bases, and dual bases (see, e.g., [BJMP05, DR08] and the papers cited there), the ONBs are the most agreeable at least from the mathematical point of view.

That there are bases of this kind is not at all clear, and the subject of wavelets in this continuous context has gained much from its connections to the discrete world of signals and image processing.

Here, we shall outline some of these connections with an emphasis on the mathematical context. So we will be stressing the theory of Hilbert space, and bounded linear operators acting in Hilbert space $\mathcal{H}$, both individual operators, and families of operators which form algebras.

As was noticed recently the operators which specify particular subband algorithms from the discrete world of signal processing turn out to satisfy relations that were found (or rediscovered independently) in the theory of operator algebras, and which go under the name of Cuntz algebras, denoted $\mathcal{O}_N$ if n is the number of bands. For additional details, see, e.g., [Jor06a].

In symbols, the C^*algebra has generators $(S_i)_{i=0}^{N-1}$, and the relations are

$$\sum_{i=0}^{N-1} S_i S_i^* = 1 \qquad (4.1)$$

(where $\mathbf{1}$ is the identity element in $\mathcal{O}_N$) and

$$\sum_{i=0}^{N-1} S_i S_i^* = \mathbf{1}, \quad \text{and} \quad S_i^* S_j = \delta_{i,j}\mathbf{1}. \qquad (4.2)$$

Please see Fig. 4.1.

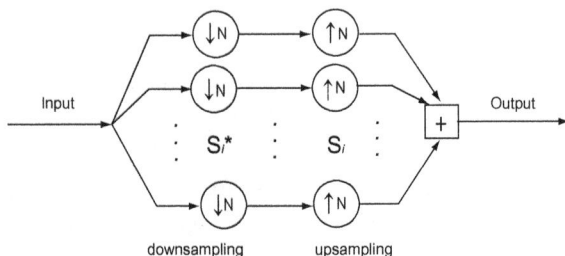

Fig. 4.1 Input and output sketch for multifrequency band signal processing: An input signal is filtered (with a separate filter for each band), then down-sampled, transmitted over separate frequency bands, upsampled, run through a dual filter system, and finally added together into a single output.

In a representation on a Hilbert space, say $\mathcal{H}$, the symbols S_i turn into bounded operators, also denoted S_i, and the identity element $\mathbf{1}$ turns into the identity operator I in $\mathcal{H}$, i.e., the operator $I : h \to h$, for $h \in \mathcal{H}$. In operator language, the two formulas (4.1) and (4.2) state that each S_i is an isometry in $\mathcal{H}$, and that the respective ranges $S_i\mathcal{H}$ are mutually orthogonal, i.e., $S_i\mathcal{H} \perp S_j\mathcal{H}$ for $i \neq j$. Introducing the projections $P_i = S_iS_i^*$, we get $P_iP_j = \delta_{i,j}P_i$, and

$$\sum_{i=0}^{N-1} P_i = I.$$

In the engineering literature, this takes the form of programming diagrams:

If the process of Figure 4.3, is repeated, we arrive at the discrete wavelet transform, or stated in the form of images ($n = 5$). This division algorithm can also be illustrated as Fig. 4.2.

Selecting a resolution subspace $V_0 = \text{closure span}\{\varphi(\cdot - k)|k \in \mathbb{Z}\}$, we arrive at a wavelet subdivision $\{\psi_{j,k}|j \geq 0, k \in \mathbb{Z}\}$, where $\psi_{j,k}(x) = 2^{j/2}\psi(2^jx - k)$, and the continuous expansion $f = \sum_{j,k}\langle\psi_{j,k}|f\rangle\psi_{j,k}$ or the discrete analogue derived from the isometries, $i = 1, 2, \ldots, N - 1$, $S_0^kS_i$ for $k = 0, 1, 2, \ldots$; are called the discrete wavelet transform.

Notational Convention: In algorithms, the letter N is popular, and often used for counting more than one thing.

In the present contest of the Discrete Wavelet Algorithm (DWA) or DWT, we count two things, "the number of times a picture is decomposed via subdivision." We have used n for this. The other related but different number N is the number of subbands, $N = 2$ for the dyadic DWT, and $N = 4$ for the image DWT. The image-processing WT in our present context is

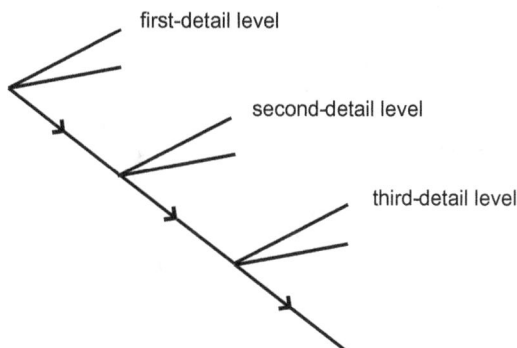

Fig. 4.2 Perfect reconstruction in a subband filtering as used in signal and image processing.

the tensor product of the 1-D dyadic WT, so $2 \times 2 = 4$. Caution: Not all DWAs arise as tensor products of $N = 2$ models. The wavelets coming from tensor products are called separable. When a particular image-processing scheme is used for generating continuous wavelets, it is not transparent if we are looking at a separable or inseparable wavelet! [JS14a].

To clarify the distinction, it is helpful to look at the representations of the Cuntz relations by operators in Hilbert space. We are dealing with representations of the two distinct algebras $\mathcal{O}_2$, and $\mathcal{O}_4$; two frequency subbands vs. 4 subbands. Note that the Cuntz $\mathcal{O}_2$, and $\mathcal{O}_4$ are given axiomatically, or purely symbolically. It is only when subband filters are chosen that we get representations. This also means that the choice of N is made initially; and the same N is used in different runs of the programs. In contrast, the number of times a picture is decomposed varies from one experiment to the next!

Summary: $N = 2$ for the dyadic DWT: The operators in the representation are S_0, S_1. One average operator, and one detail operator. The detail operator S_1 "counts" local detail variations [JS14a].

Image-processing: Then $N = 4$ is fixed as we run different images in the DWT: The operators are now: S_0 , S_H, S_V, S_D. One average operator, and three detail operators for local detail variations in the three directions in the plane. An image illustration example would be Fig. 4.4. The image has been inverted in color for better view.

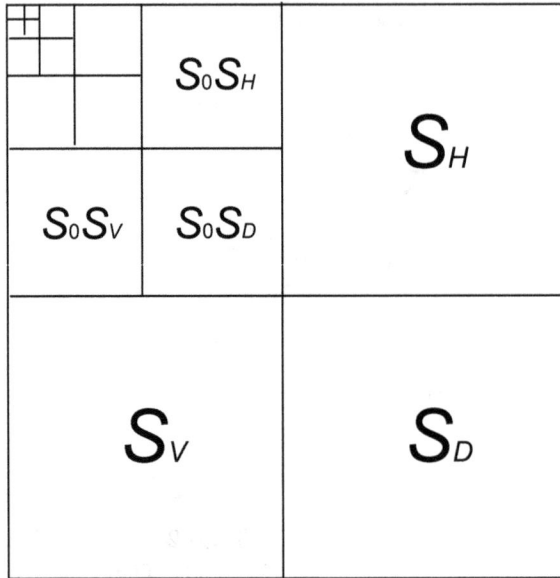

Fig. 4.3 The subdivided squares represent the use of the pyramid subdivision algorithm to image processing, as it is used on pixel squares. At each subdivision step the top left-hand square represents averages of nearby pixel numbers, averages taken with respect to the chosen low-pass filter; while the three directions, horizontal, vertical, and diagonal, represent detail differences, with the three represented by separate bands and filters. So in this model, there are four bands, and they may be realized by a tensor product construction applied to dyadic filters in the separate x- and the y-directions in the plane. For the discrete WT used in image-processing, we use iteration of four isometries S_0, S_H, S_V, and S_D with mutually orthogonal ranges, and satisfying the following sum-rule $S_0 S_0^* + S_H S_H^* + S_V S_V^* + S_D S_D^* = I$, with I denoting the identity operator in an appropriate l^2-space.

4.3 The Continuous Wavelet Transform

Consider functions f on the real line $\mathbb{R}$. We select the Hilbert space of functions to be $L^2(\mathbb{R})$. To start a continuous WT, we must select a function $\psi \in L^2(\mathbb{R})$ and $r, s \in \mathbb{R}$ such that the following family of functions

$$\psi_{r,s}(x) = r^{-1/2} \psi \left(\frac{x-s}{r} \right),$$

creates an over-complete basis for $L^2(\mathbb{R})$. An over-complete family of vectors in a Hilbert space is often called a coherent decomposition. This terminology comes from quantum optics. What is needed for a continuous WT in the

Fig. 4.4 $n = 2$ Jorgensen.

simplest case is the following representation valid for all $f \in L^2(\mathbb{R})$:

$$f(x) = C_\psi^{-1} \int \int_{\mathbb{R}^2} \langle \psi_{r,s} | f \rangle \psi_{r,s}(x) \frac{dr\, ds}{r^2},$$

where $C_\psi := \int_{\mathbb{R}} |\hat{\psi}(\omega)|^2 \frac{d\omega}{\omega}$ and where $\langle \psi_{r,s} | f \rangle = \int_{\mathbb{R}} \overline{\psi_{r,s}(y)} f(y) dy$. The refinements and implications of this are spelled out in tables in Section 4.4.

4.3.1 *Some background on Hilbert space*

Wavelet theory is the art of finding a special kind of basis in Hilbert space. Let $\mathcal{H}$ be a Hilbert space over $\mathbb{C}$ and denote the inner product $\langle \cdot \mid \cdot \rangle$. For us, it is assumed linear in the second variable. If $\mathcal{H} = L^2(\mathbb{R})$, then

$$\langle f \mid g \rangle := \int_{\mathbb{R}} \overline{f(x)} \, g(x) \, dx.$$

If $\mathcal{H} = \ell^2(\mathbb{Z})$, then

$$\langle \xi \mid \eta \rangle := \sum_{n \in \mathbb{Z}} \bar{\xi}_n \eta_n.$$

Let $\mathbb{T} = \mathbb{R}/2\pi\mathbb{Z}$. If $\mathcal{H} = L^2(\mathbb{T})$, then

$$\langle f \mid g \rangle := \frac{1}{2\pi} \int_{-\pi}^{\pi} \overline{f(\theta)} \, g(\theta) \, d\theta.$$

Functions $f \in L^2(\mathbb{T})$ have Fourier series: Setting $e_n(\theta) = e^{in\theta}$,

$$\hat{f}(n) := \langle e_n \mid f \rangle = \frac{1}{2\pi} \int_{-\pi}^{\pi} e^{-in\theta} f(\theta) \, d\theta,$$

and

$$\|f\|_{L^2(\mathbb{T})}^2 = \sum_{n \in \mathbb{Z}} \left| \hat{f}(n) \right|^2.$$

Similarly, if $f \in L^2(\mathbb{R})$, then

$$\hat{f}(t) := \int_{\mathbb{R}} e^{-ixt} f(x) \, dx,$$

and

$$\|f\|_{L^2(\mathbb{R})}^2 = \frac{1}{2\pi} \int_{\mathbb{R}} \left| \hat{f}(t) \right|^2 \, dt.$$

Let J be an index set. We shall only need to consider the case when J is countable. Let $\{\psi_\alpha\}_{\alpha \in J}$ be a family of nonzero vectors in a Hilbert space $\mathcal{H}$. We say it is an *orthonormal basis* (ONB) if

$$\langle \psi_\alpha \mid \psi_\beta \rangle = \delta_{\alpha,\beta} \qquad \text{(Kronecker delta)} \qquad (4.3)$$

and if

$$\sum_{\alpha \in J} |\langle \psi_\alpha \mid f \rangle|^2 = \|f\|^2 \quad \text{holds for all } f \in \mathcal{H}. \qquad (4.4)$$

If only (4.4) is assumed, but not (4.3), we say that $\{\psi_\alpha\}_{\alpha \in J}$ is a (normalized) *tight frame*. We say that it is a *frame* with *frame constants*

$0 < A \le B < \infty$ if

$$A \, \|f\|^2 \le \sum_{\alpha \in J} |\langle \, \psi_\alpha \mid f \, \rangle|^2 \le B \, \|f\|^2 \qquad \text{holds for all } f \in \mathcal{H}.$$

Introducing the rank-one operators $Q_\alpha := |\psi_\alpha\rangle \langle\psi_\alpha|$ of Dirac's terminology, see [BJ02a], we see that $\{\psi_\alpha\}_{\alpha \in J}$ is an ONB if and only if the Q_α's are projections, and

$$\sum_{\alpha \in J} Q_\alpha = I \qquad (= \text{the identity operator in } \mathcal{H}). \qquad (4.5)$$

It is a (normalized) tight frame if and only if (4.5) holds but with no further restriction on the rank-one operators Q_α. It is a frame with frame constants A and B if the operator

$$S := \sum_{\alpha \in J} Q_\alpha$$

satisfies

$$AI \le S \le BI$$

in the order of Hermitian operators. (We say that operators $H_i = H_i^*$, $i = 1, 2$, satisfy $H_1 \le H_2$ if $\langle \, f \mid H_1 f \, \rangle \le \langle \, f \mid H_2 f \, \rangle$ holds for all $f \in \mathcal{H}$). If h, k are vectors in a Hilbert space $\mathcal{H}$, then the operator $A = |h\rangle \langle k|$ is defined by the identity $\langle \, u \mid Av \, \rangle = \langle \, u \mid h \, \rangle \langle \, k \mid v \, \rangle$ for all $u, v \in \mathcal{H}$.

Wavelets in $L^2(\mathbb{R})$ are generated by simple operations on one or more functions ψ in $L^2(\mathbb{R})$, the operations come in pairs, say scaling and translation, or phase-modulation and translations. If $N \in \{2, 3, \dots\}$, we set

$$\psi_{j,k}(x) := N^{j/2} \psi \left(N^j x - k \right) \quad \text{for } j, k \in \mathbb{Z}.$$

4.3.1.1 *Increasing the dimension*

In wavelet theory [Dau92], there is a tradition for reserving φ for the father function and ψ for the mother function. A 1-level wavelet transform of an

$N \times M$ image can be represented as

$$\mathbf{f} \mapsto \begin{pmatrix} \mathbf{a}^1 & | & \mathbf{h}^1 \\ -- & & -- \\ \mathbf{v}^1 & | & \mathbf{d}^1 \end{pmatrix}, \tag{4.6}$$

where the subimages $\mathbf{h}^1, \mathbf{d}^1, \mathbf{a}^1$, and $\mathbf{v}^1$ each have the dimension of $N/2$ by $M/2$.

$$\mathbf{a}^1 = V_m^1 \otimes V_n^1 : \varphi^A(x,y) = \varphi(x)\varphi(y) = \sum_i \sum_j h_i h_j \varphi(2x-i)\varphi(2y-j)$$
$$\mathbf{h}^1 = V_m^1 \otimes W_n^1 : \psi^H(x,y) = \psi(x)\varphi(y) = \sum_i \sum_j g_i h_j \varphi(2x-i)\varphi(2y-j)$$
$$\mathbf{v}^1 = W_m^1 \otimes V_n^1 : \psi^V(x,y) = \varphi(x)\psi(y) = \sum_i \sum_j h_i g_j \varphi(2x-i)\varphi(2y-j)$$
$$\mathbf{d}^1 = W_m^1 \otimes W_n^1 : \psi^D(x,y) = \psi(x)\psi(y) = \sum_i \sum_j g_i g_j \varphi(2x-i)\varphi(2y-j), \tag{4.7}$$

where φ is the father function, and ψ is the mother function in the sense of wavelet, V space, denotes the average space, and the W spaces are the difference spaces from multiresolution analysis (MRA) [Dau92].

In the formulas, we have the following two indexed number systems $\mathbf{a} := (h_i)$ and $\mathbf{d} := (g_i)$, $\mathbf{a}$ is for averages and $\mathbf{d}$ is for local differences. They are really the input for the DWT. But they also form the key link between the two transforms, the discrete and continuous. The link is made up of the following scaling identities:

$$\varphi(x) = 2 \sum_{i \in \mathbb{Z}} h_i \varphi(2x-i);$$

$$\psi(x) = 2 \sum_{i \in \mathbb{Z}} g_i \varphi(2x-i);$$

and (low-pass normalization) $\sum_{i \in \mathbb{Z}} h_i = 1$. The scalars (h_i) may be real or complex; they may be finite or infinite in number. If there are four of them, it is called the "four tap," etc. The finite case is best for computations since it corresponds to compactly supported functions. This means that the two functions φ and ψ will vanish outside some finite interval on a real line.

The two number systems are further subjected to orthogonality relations, of which

$$\sum_{i \in \mathbb{Z}} \bar{h}_i h_{i+2k} = \frac{1}{2} \delta_{0,k} \tag{4.8}$$

is the best known.

The systems h and g are both low-pass and high-pass filter coefficients. In (4.7), $\mathbf{a}^1$ denotes the first averaged image, which consists of average intensity values of the original image. Note that only φ function, V space and h coefficients are used here. Similarly, $\mathbf{h}^1$ denotes the first detail image of horizontal components, which consists of intensity difference along the vertical axis of the original image. Note that φ function is used on y and ψ function on x, W space for x values and V space for y values; and both h and g coefficients are used accordingly. The data $\mathbf{v}^1$ denotes the first detail image of vertical components, which consists of intensity difference along the horizontal axis of the original image. Note that φ function is used on x and ψ function on y; W space for y values and V space for x values; and both h and g coefficients are used accordingly. Finally, $\mathbf{d}^1$ denotes the first detail image of diagonal components, which consists of intensity difference along the diagonal axis of the original image. The original image is reconstructed from the decomposed image by taking the sum of the averaged image and the detail images and scaling by a scaling factor. It could be noted that only ψ function, W space, and g coefficients are used here. See [Wal99b], [Son06b].

This decomposition not only limits to one step, but it can be done again and again on the averaged detail depending on the size of the image. Once it stops at a certain level, quantization (see [SCE01, Use01]) is done on the image. This quantization step may be lossy or lossless. Then the lossless entropy encoding is done on the decomposed and quantized image.

The relevance of the system of identities (4.8) may be summarized as follows. Set

$$m_0(z) := \frac{1}{2}\sum_{k\in\mathbb{Z}} h_k z^k \quad \text{for all } z \in \mathbb{T};$$

$$g_k := (-1)^k \bar{h}_{1-k} \quad \text{for all } k \in \mathbb{Z};$$

$$m_1(z) := \frac{1}{2}\sum_{k\in\mathbb{Z}} g_k z^k; \quad \text{and}$$

$$S_j f)(z) = \sqrt{2} m_j(z) f(z^2), \text{ for } j = 0, 1, \ f \in L^2(\mathbb{T}), \ z \in \mathbb{T}.$$

Then the following conditions are equivalent:

(a) The system of Equations (4.8) is satisfied.
(b) The operators S_0 and S_1 satisfy the Cuntz relations.
(c) We have perfect reconstruction in the subband system of Figure 4.3.

Note that the two operators S_0 and S_1 have equivalent matrix representations. Recall that by Parseval's formula, we have $L^2(\mathbb{T}) \simeq l^2(\mathbb{Z})$. So representing S_0 instead as an $\infty \times \infty$ matrix acting on column vectors $x = (x_j)_{j \in \mathbb{Z}}$, we get

$$(S_0 x)_i = \sqrt{2} \sum_{j \in \mathbb{Z}} h_{i-2j} x_j$$

and for the adjoint operator $F_0 := S_0^*$, we get the matrix representation

$$(F_0 x)_i = \frac{1}{\sqrt{2}} \sum_{j \in \mathbb{Z}} \bar{h}_{j-2i} x_j$$

with the overbar signifying complex conjugation. This is of computational significance to the two matrix representations, both the matrix for S_0 and for $F_0 := S_0^*$ are slanted. However, the slanting of one is the mirror-image of the other, i.e.,

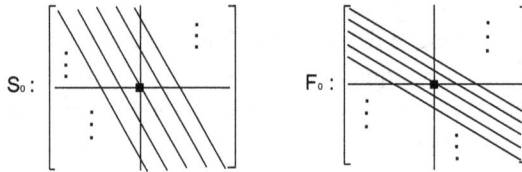

Significance of slanting. The slanted matrix representations refer to the corresponding operators in L^2. In general, operators in Hilbert function spaces have many matrix representations, one for each orthonormal basis (ONB), but here we are concerned with the ONB consisting of the Fourier frequencies z^j, $j \in \mathbb{Z}$. So in our matrix representations for the S operators and their adjoints, we will be acting on column vectors, each infinite column representing a vector in the sequence space l^2. A vector in l^2 is said to be of finite size if it has only a finite set of nonzero entries.

It is the matrix F_0 that is effective for iterated matrix computation. Reason: When a column vector x of a fixed size, say 2 s, is multiplied, or acted on by F_0, the result is a vector y of half the size, i.e., of size s. So $y = F_0 x$. If we use F_0 and F_1 together on x, then we get two vectors, each of size s, the other one $z = F_1 x$, and we can form the combined column vector of y and z; stacking y on top of z. In our application, y represents averages, while z represents local differences: hence, the wavelet algorithm [JS14a].

$$\begin{bmatrix} \vdots \\ y_{-1} \\ y_0 \\ y_1 \\ \vdots \\ -- \\ \vdots \\ z_{-1} \\ z_0 \\ z_1 \\ \vdots \end{bmatrix} = \begin{bmatrix} F_0 \\ -- \\ F_1 \end{bmatrix} \begin{bmatrix} \vdots \\ x_{-2} \\ x_{-1} \\ x_0 \\ x_1 \\ x_2 \\ \vdots \end{bmatrix}$$

$$y = F_0 x$$
$$z = F_1 x.$$

4.4 Connections to Group Theory

The first line in the two following tables is the continuous wavelet transform. It comes from what in physics is called *coherent vector decompositions*. Both transforms apply to vectors in Hilbert space $\mathcal{H}$, and $\mathcal{H}$ may vary from case to case. Common to all transforms is vector input and output. If the input agrees with output, we say that the combined process yields the identity operator image [JS14a]. $\mathbf{1} : \mathcal{H} \to \mathcal{H}$ or written $\mathbf{1}_{\mathcal{H}}$. So, e.g., if $(S_i)_{i=0}^{N-1}$ is a finite operator system, input/output operator example may take the form

$$\sum_{i=0}^{N-1} S_i S_i^* = \mathbf{1}_{\mathcal{H}}.$$

Summary of and variations on the resolution of the identity operator $\mathbf{1}$ in L^2 or in ℓ^2, for ψ and $\tilde{\psi}$ where $\psi_{r,s}(x) = r^{-\frac{1}{2}} \psi\left(\frac{x-s}{r}\right)$,

$$C_\psi = \int_{\mathbb{R}} \frac{d\omega}{|\omega|} |\hat{\psi}(\omega)|^2 < \infty,$$

similarly for $\tilde{\psi}$ and $C_{\psi,\tilde{\psi}} = \int_{\mathbb{R}} \frac{d\omega}{|\omega|} \overline{\hat{\psi}(\omega)} \hat{\tilde{\psi}}(\omega)$:

$N = 2$	**Over-complete Basis**	**Dual Bases**
Continuous resolution	$C_\psi^{-1} \iint\limits_{\mathbb{R}^2} \dfrac{dr\,ds}{r^2}\,\lvert\psi_{r,s}\rangle\langle\psi_{r,s}\rvert = 1$	$C_{\psi,\tilde\psi}^{-1} \iint\limits_{\mathbb{R}^2} \dfrac{dr\,ds}{r^2}\,\lvert\psi_{r,s}\rangle\langle\tilde\psi_{r,s}\rvert$
		$= 1$
Discrete resolution	$\displaystyle\sum_{j\in\mathbb{Z}}\sum_{k\in\mathbb{Z}}\lvert\psi_{j,k}\rangle\langle\psi_{j,k}\rvert = 1,$	$\displaystyle\sum_{j\in\mathbb{Z}}\sum_{k\in\mathbb{Z}}\lvert\psi_{j,k}\rangle\langle\tilde\psi_{j,k}\rvert = 1$
	$\psi_{j,k}$ corresponding to $r = 2^{-j}$, $s = k2^{-j}$	

$N \geq 2$	**Isometries in ℓ^2**	**Dual Operator System in ℓ^2**
Sequence spaces	$\displaystyle\sum_{i=0}^{N-1} S_i S_i^* = 1,$	$\displaystyle\sum_{i=0}^{N-1} S_i \tilde S_i^* = 1,$
	where $S_0,\dots,S_{N-1}$ are adjoints to the quadrature mirror filter operators F_i, i.e., $S_i = F_i^*$	for a dual operator system $S_0,\dots,S_{N-1}$, $\tilde S_0,\dots,\tilde S_{N-1}$

Then the assertions in the first table amount to

$$C_\psi^{-1} \iint\limits_{\mathbb{R}^2} \frac{dr\,ds}{r^2}\,\lvert\langle\psi_{r,s}\mid f\rangle\rvert^2 \qquad C_{\psi,\tilde\psi}^{-1} \iint\limits_{\mathbb{R}^2} \frac{dr\,ds}{r^2}\,\langle f\mid\psi_{r,s}\rangle\langle\tilde\psi_{r,s}\mid g\rangle$$

$$= \lVert f\rVert_{L^2}^2 \quad \forall f\in L^2(\mathbb{R}) \qquad\qquad = \langle f\mid g\rangle \quad \forall f,g\in L^2(\mathbb{R})$$

$$\sum_{j\in\mathbb{Z}}\sum_{k\in\mathbb{Z}}\lvert\langle\psi_{j,k}\mid f\rangle\rvert^2 \qquad\qquad \sum_{j\in\mathbb{Z}}\sum_{k\in\mathbb{Z}}\langle f\mid\psi_{j,k}\rangle\langle\tilde\psi_{j,k}\mid g\rangle$$

$$= \lVert f\rVert_{L^2}^2 \quad \forall f\in L^2(\mathbb{R}) \qquad\qquad = \langle f\mid g\rangle \quad \forall f,g\in L^2(\mathbb{R})$$

$$\sum_{i=0}^{N-1}\lVert S_i^* c\rVert^2 = \lVert c\rVert^2 \quad \forall c\in\ell^2 \qquad \sum_{i=0}^{N-1}\langle S_i^* c\mid\tilde S_i^* d\rangle = \langle c\mid d\rangle \quad \forall c,d\in\ell^2$$

A function ψ satisfying the resolution identity is called a *coherent vector* in mathematical physics. The representation theory for the $(ax + b)$-group, i.e., the matrix group $G = \{ \left(\begin{smallmatrix} a & b \\ 0 & 1 \end{smallmatrix} \right) \mid a \in \mathbb{R}_+, \ b \in \mathbb{R} \}$, serves as its underpinning. Then the tables above illustrate how the $\{\psi_{j,k}\}$ wavelet system arises from a discretization of the following unitary representation of G:

$$\left(U_{\left(\begin{smallmatrix} a & b \\ 0 & 1 \end{smallmatrix} \right)} f \right)(x) = a^{-\frac{1}{2}} f\left(\frac{x - b}{a} \right)$$

acting on $L^2(\mathbb{R})$. This unitary representation also explains the discretization step in passing from the first line to the second in the tables above. The functions $\{ \psi_{j,k} \mid j, k \in \mathbb{Z} \}$, which make up a wavelet system, result from the choice of a suitable coherent vector $\psi \in L^2(\mathbb{R})$, and then setting

$$\psi_{j,k}(x) = \left(U_{\left(\begin{smallmatrix} 2^{-j} & k \cdot 2^{-j} \\ 0 & 1 \end{smallmatrix} \right)} \psi \right)(x) = 2^{\frac{j}{2}} \psi \left(2^j x - k \right).$$

Even though this representation lies at the historical origin of the subject of wavelets, the $(ax + b)$-group seems to be now largely forgotten in the next generation of the wavelet community. But Chapters 1–3 of [Dau92] still serve as a beautiful presentation of this (now much ignored) side of the subject. It also serves as a link to mathematical physics and to classical analysis.

4.5 Tools from Mathematics

In our presentation, we will rely on tools from at least three separate areas of mathematics, and we will outline how they interact to form a coherent theory, and how they come together to form a link between what is now called the discrete and the continuous wavelet transform. It is the discrete case that is popular with engineers ([AK06b, Liu06, Str97, Str00]), while the continuous case has come to play a central role in the part of mathematics referred to as harmonic analysis, [Dau93]. The three areas are operator algebras, dynamical systems, and basis constructions [JS14a]:

(a) **Operator Algebras:** The theory of operator algebras in turn breaks up in two parts: One, the study of "the algebras themselves" as they emerge from the axioms of von Neumann (von Neumann algebras), and Gelfand, Kadison, and Segal (C^*algebras). The other has a more applied slant: It involves "the representations" of the algebras. By this, we refer to the following: The algebras will typically be specified by generators and by relations, and by a certain norm-completion, in any case by a system of axioms. This holds both for the norm-closed algebras, the so-called C^* algebras, and for the weakly closed algebras, the von Neumann algebras. In fact, there is a close connection between the two parts of the theory: For example, representations of C^* algebras generate von Neumann algebras.

To talk about representations of a fixed algebra, say A, we must specify a Hilbert space, and a homomorphism ρ from A into the algebra $\mathcal{B}(H)$ of all bounded operators on $\mathcal{H}$. We require that ρ sends the identity element in A into the identity operator acting on $\mathcal{H}$, and that $\rho(a^*) = (\rho(a))^*$ where the last star now refers to the adjoint operator.

It was realized in the last 10 years (see, e.g., [BJ02a, Jor06a, Jor06b] that a family of representations of wavelets which are basis constructions in harmonic analysis, in signal/image analysis, and in computational mathematics, may be built up from representations of an especially important family of simple C^* algebras, the Cuntz algebras. The Cuntz algebras are denoted $\mathcal{O}_2, \mathcal{O}_3, ...$, including $\mathcal{O}_\infty$.

(b) **Dynamical systems:** The connection between the Cuntz algebras $\mathcal{O}_N$ for $N = 2, 3, \ldots$ are relevant to the kind of dynamical systems which are built on branching-laws, the case of $\mathcal{O}_N$ representing N-fold branching. The reason for this is that if N is fixed, $\mathcal{O}_N$ includes in its definition an iterated subdivision, but within the context of Hilbert space. For more details, see, e.g., [Dut04, DR07, DJ05, DJ06b, DJ06c, DJ06a, Jor06b]. Readers are referred to Appendix C for basics on dynamical systems and Cantor dynamics.

(c) **Analysis of bases in function spaces:** The connection to basis constructions using wavelets is this: The context for wavelets is a Hilbert space $\mathcal{H}$, where $\mathcal{H}$ may be $L^2(\mathbb{R}^d)$ where d is a dimension, $d = 1$ for the line (signals), $d = 2$ for the plane (images), etc. The more successful bases in Hilbert space are the orthonormal bases ONBs, but until the mid 1980s, there were no ONBs in $L^2(\mathbb{R}^d)$ which were entirely algorithmic and effective for computations. One reason for this is that the tools that had been used for 200 years since Fourier involved basis functions (Fourier wave functions), which were not localized. Moreover these existing Fourier tools were not friendly to algorithmic computations.

4.6 A Transfer Operator

A popular tool for deciding if a candidate for a wavelet basis is in fact an ONB uses a certain transfer operator. Variants of this operator are used in diverse areas of applied mathematics. It is an operator which involves a weighted average over a finite set of possibilities. Hence, it is natural for understanding random walk algorithms. As remarked in, e.g., [Jor03, Jor06a, Jor06b, Dut04], it was also studied in physics, e.g., by David Ruelle who used to prove results on phase transition for infinite spin systems in quantum statistical mechanics. In fact the transfer operator has many incarnations (many of them known as Ruelle operators), and all of them based on N-fold branching laws.

In our wavelet application, the Ruelle operator weights in input over the N branch possibilities, and the weighting is assigned by a chosen scalar function W. The W-Ruelle operator is denoted R_W. In the wavelet setting there is in addition a low-pass filter function m_0, which in its frequency response formulation is a function on the d-torus $\mathbf{T}^d = \mathbb{R}^d/\mathbb{Z}^d$.

Since the scaling matrix A has integer entries, A passes to the quotient $\mathbb{R}^d/\mathbb{Z}^d$, and the induced transformation $r_A : \mathbb{T}^d \to \mathbb{T}^d$ is an N-fold cover, where $N = |\det A|$, i.e., for every x in $\mathbb{T}^d$ there are N distinct points y in $\mathbb{T}^d$ solving $r_A(y) = x$.

In the wavelet case, the weight function W is $W = |m_0|^2$. Then with this choice of W, the ONB problem for a candidate for a wavelet basis in the Hilbert space $L^2(\mathbb{R}^d)$ as it turns out may be decided by the dimension of a distinguished eigenspace for R_W, by the so-called Perron–Frobenius problem.

This has worked well for years for the wavelets which have an especially simple algorithm, the wavelets that are initialized by a single function, called the scaling function. These are called the multiresolution analysis (MRA) wavelets, or for short the MRA-wavelets. But there are instances, e.g., if a problem must be localized in frequency domain, when the MRA-wavelets do not suffice, where it will by necessity include more than one scaling function. And we are then back to trying to decide if the output from the discrete algorithm and the $\mathcal{O}_N$ representation is an ONB, or if it has some stability property which will serve the same purpose, in case where asking for an ONB is not feasible.

4.7 Future Directions

The idea of a scientific analysis by subdividing a fixed picture or object into its finer parts is not unique to wavelets. It works best for structures with an inherent self-similarity; this self-similarity can arise from numerical scaling of distances. But there are more subtle nonlinear self-similarities. The Julia sets in the complex plane are a case in point [BY06, Bra06, DL06, DRS07, Mil04, PZ04]. The simplest Julia set comes from a one parameter family of quadratic polynomials $\varphi_c(z) = z^2 + c$, where z is a complex variable and where c is a fixed parameter. The corresponding Julia sets J_c have a surprisingly rich structure. A simple way to understand them is the following: Consider the two branches of the inverse $\beta_\pm = z \mapsto \pm\sqrt{z-c}$. Then J_c is the unique minimal non-empty compact subset of $\mathbb{C}$, which is invariant under $\{\beta_\pm\}$. (There are alternative ways of presenting J_c, but this one fits our purpose. The Julia set J of a holomorphic function, in this case $z \mapsto z^2 + c$ (see Figs. 4.5 and 4.6), informally consists of those points whose long-time behavior under repeated iteration, or rather iteration of substitutions, can change drastically under arbitrarily small perturbations.) Here, "long-time" refers to large n, where $\varphi^{(n+1)}(z) = \varphi(\varphi^{(n)}(z))$, $n = 0, 1, ...$, and $\varphi^{(0)}(z) = z$.

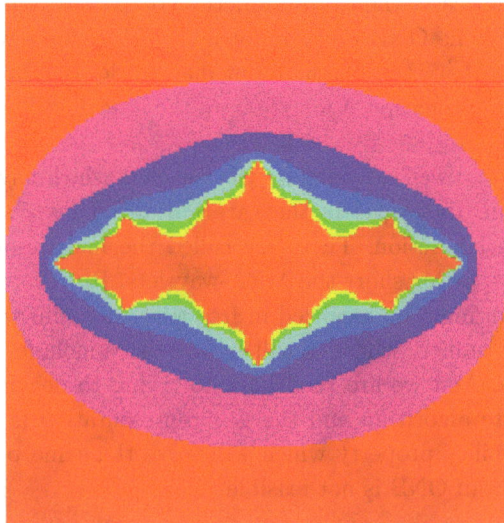

Fig. 4.5 Julia set with $c = -1$.

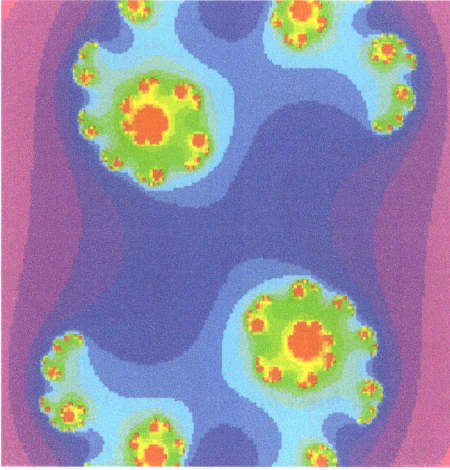

Fig. 4.6 Julia set with $c = 0.45 - 0.1428i$.

It would be interesting to adapt and modify the Haar wavelet, and the other wavelet algorithms to the Julia sets. The two papers [DJ05, DJ06b] initiated such a development. Then an attempt to adapt and modify the Haar wavelet to the Julia sets was made [DPS14], however, there were some limitations in finding the filters. Perhaps trying another fractal set such as tent map or others may work.

4.7.1 *Orthonormal bases generated by Cuntz algebras*

We present new results from [DPS14] by borrowing Section 3 and part of Section 2 from [DPS14] in the rest of this Section 4.7.1. It gives a general criterion for a family generated by the Cuntz isometries to be an orthonormal basis.

Theorem 4.1. [DPS14] *Let $\mathcal{H}$ be a Hilbert space and $(S_i)_{i=0}^{N-1}$ be a representation of the Cuntz algebra $\mathcal{O}_N$. Let $\mathcal{E}$ be an orthonormal set in $\mathcal{H}$ and $f : X \to \mathcal{H}$ a norm continuous function on a topological space X with the following properties:*

(1) $\mathcal{E} = \cup_{i=0}^{N-1} S_i \mathcal{E}$.
(2) $\overline{\text{span}}\{f(t) : t \in X\} = \mathcal{H}$ and $\|f(t)\| = 1$, for all $t \in X$.

(3) *There exist functions* $\mathfrak{m}_i : X \to \mathbb{C}$, $g_i : X \to X$, $i = 0, \ldots, N-1$ *such that*

$$S_i^* f(t) = \mathfrak{m}_i(t) f(g_i(t)), \quad t \in X. \tag{4.9}$$

(4) *There exist* $c_0 \in X$ *such that* $f(c_0) \in \overline{\text{span}}\mathcal{E}$.

(5) *The only functions* $h \in C(X)$ *with* $h \geq 0$, $h(c) = 1$, $\forall\, c \in \{x \in X : f(x) \in \overline{\text{span}}\mathcal{E}\}$, *and*

$$h(t) = \sum_{i=0}^{N-1} |\mathfrak{m}_i(t)|^2 h(g_i(t)), \quad t \in X \tag{4.10}$$

are the constant functions.

Then $\mathcal{E}$ *is an orthonormal basis for* $\mathcal{H}$.

Proof. Define $h(t) := \sum_{e \in \mathcal{E}} |\langle f(t), e \rangle|^2 = \|Pf(t)\|^2$, $t \in X$ where P is the orthogonal projection onto the closed linear span of $\mathcal{E}$.

Since $t \mapsto f(t)$ is norm continuous, we get that h is continuous. Clearly, $h \geq 0$. Also, if $f(c) \in \overline{\text{span}}\mathcal{E}$, then $\|Pf(c)\| = \|f(c)\| = 1$ so $h(c) = 1$. In particular, from (ii) and (iv), $h(c_0) = 1$. We check (4.10). Since the sets $S_i \mathcal{E}$, $i = 0, \ldots, N-1$ are mutually orthogonal, the union in (i) is disjoint. Therefore, for all $t \in X$:

$$h(t) = \sum_{i=0}^{N-1} \sum_{e \in \mathcal{E}} |\langle f(t), S_i e \rangle|^2 = \sum_{i=0}^{N-1} \sum_{e \in \mathcal{E}} |\langle S_i^* f(t), e \rangle|^2$$

$$= \sum_{i=0}^{N-1} |\mathfrak{m}_i(t)|^2 \sum_{e \in \mathcal{E}} |\langle f(g_i(t)), e \rangle|^2$$

$$= \sum_{i=0}^{N-1} |\mathfrak{m}_i(t)|^2 h(g_i(t)).$$

By (v), h is constant and, since $h(c_0) = 1$, $h(t) = 1$ for all $t \in X$. Then $\|Pf(t)\| = 1$ for all $t \in X$. Since $\|f(t)\| = 1$, it follows that $f(t) \in \text{span}\mathcal{E}$ for all $t \in X$. But the vectors $f(t)$ span $\mathcal{H}$ so $\overline{\text{span}}\mathcal{E} = \mathcal{H}$ and $\mathcal{E}$ is an orthonormal basis. $\qquad\qquad\square$

Remark 4.2. [DPS14] The operators of the form

$$Rh(t) = \sum_{i=0}^{N-1} |\mathfrak{m}_i(t)|^2 h(g_i(t)), \quad t \in X, h \in C(X),$$

that appear in (4.10), are sometimes called Ruelle operators or transfer operators, see, e.g., [Bal00].

4.7.1.1 *Piecewise exponential bases on fractals*

Example 4.3. [DPS14] We consider affine-iterated function systems with no overlap. Let R be a $d \times d$ expansive real matrix, i.e., all the eigenvalues of R have absolute value strictly greater than 1. Let $B \subset \mathbb{R}^d$ a finite set such that $N = |B|$. Define the affine-iterated function system

$$\tau_b(x) = R^{-1}(x + b) \quad (x \in \mathbb{R}^d, \, b \in B). \tag{4.11}$$

By [Hut81], there exists a unique compact subset X_B of $\mathbb{R}^d$ which satisfies the invariance equation

$$X_B = \cup_{b \in B} \tau_b(X_B). \tag{4.12}$$

X_B is called the attractor of the iterated function system $(\tau_b)_{b \in B}$. Moreover, X_B is given by

$$X_B = \left\{ \sum_{k=1}^{\infty} R^{-k} b_k : b_k \in B \text{ for all } k \geq 1 \right\}. \tag{4.13}$$

Also from [Hut81], there is a unique probability measure μ_B on $\mathbb{R}^d$ satisfying the invariance equation

$$\int f d\mu_B = \frac{1}{N} \sum_{b \in B} \int f \circ \tau_b d\mu_B \tag{4.14}$$

for all continuous compactly supported functions f on $\mathbb{R}$. We call μ_B the invariant measure for the iterated function system (IFS) $(\tau_b)_{b \in B}$. By [Hut81], μ_B is supported on the attractor X_B. We say that the IFS has no overlap if $\mu_B(\tau_b(X_B) \cap \tau_{b'}(X_B)) = \emptyset$ for all $b \neq b'$ in B.

Assume that the IFS $(\tau_b)_{b \in B}$ has no overlap. Define the map $r : X_B \to X_B$

$$r(x) = \tau_b^{-1}(x), \text{ if } x \in \tau_b(X_B). \tag{4.15}$$

Then r is an N-to-1 onto map and μ_B is strongly invariant for r. Note that $r^{-1}(x) = \{\tau_b(x) : b \in B\}$ for μ_B. a.e. $x \in X_B$.

We apply Theorem 4.1 to the setting of Example 4.3, in dimension $d = 1$ for affine-iterated function systems, when the set $\frac{1}{R}B$ has a spectrum L [DPS14].

Definition 4.4. [DPS14] Let L in $\mathbb{R}$, $|L| = N$, $R > 1$ such that L is a spectrum for the set $\frac{1}{R}B$. We say that $c \in \mathbb{R}$ is an *extreme cycle point* for (B, L) if there exists $l_0, l_1, \ldots, l_{p-1}$ in L such that if $c_0 = c$, $c_1 = \frac{c_0+l_0}{R}$, $c_2 = \frac{c_1+l_1}{R}, \ldots, c_{p-1} = \frac{c_{p-2}+l_{p-2}}{R}$, then $\frac{c_{p-1}+l_{p-1}}{R} = c_0$, and $|m_B(c_i)| = 1$ for $i = 0, \ldots, p-1$ where

$$m_B(x) = \frac{1}{N} \sum_{b \in B} e^{2\pi i bx} \quad x \in \mathbb{R}.$$

Proposition 4.5. *[DPS14] Let $(m_i)_{i=0}^{N-1}$ be a QMF basis. Define the operators on $L^2(X, \mu)$*

$$S_i(f) = m_i f \circ r, \quad i = 0, \ldots, N-1. \tag{4.16}$$

Then the operators S_i are isometries and they form a representation of the Cuntz algebra $\mathcal{O}_N$, i.e.,

$$S_i^* S_j = \delta_{ij}, \quad i,j = 0, \ldots, N-1, \quad \sum_{i=0}^{N-1} S_i S_i^* = I. \tag{4.17}$$

The adjoint of S_i is given by the formula

$$S_i^*(f)(z) = \frac{1}{N} \sum_{r(w)=z} \overline{m_i}(w) f(w). \tag{4.18}$$

Proof. We compute the adjoint: take f, g in $L^2(X, \mu)$. We use the strong invariance of μ.

$$\langle S_i^* f, g \rangle = \int f \overline{m_i} \overline{g \circ r} \, d\mu = \int \frac{1}{N} \sum_{r(w)=z} \overline{m_i}(w) f(w) \overline{g}(z) d\mu(z).$$

Then (4.18) follows. The Cuntz relations in (4.17) are then easily checked with Proposition 4.6. □

Proposition 4.6. [DPS14] *A set of functions $(m_i)_{i=0}^{N-1}$ in $L^\infty(X, \mu)$ is a QMF basis if and only if*

$$\mathbb{E}(m_i \overline{m_j}) = \delta_{ij}, \quad (i,j \in \{0, \ldots, N-1\}). \tag{4.19}$$

In this case, any function $f \in L^2(X, \mu)$ can be written in the QMF basis as

$$f = \sum_{i=0}^{N-1} \mathbb{E}(f \overline{m_i}) m_i. \tag{4.20}$$

Proof. The first statement is clear. For the second, define for $f \in L^2(X, \mu)$ the vector-valued function $F(f)(z) = (f(w))_{r(w)=r(z)} \in \mathbb{C}^N$. Note that the QMF basis property implies that $(F(\frac{1}{\sqrt{N}}m_i)(z))_{i=0}^{N-1}$ is an orthonormal basis in $\mathbb{C}^n$. Then for $z \in X$

$$F(f)(z) = \sum_{i=0}^{N-1} \left\langle F(f)(z), F(\frac{1}{\sqrt{N}}m_i)(z) \right\rangle_{\mathbb{C}^N}$$

$$F\left(\frac{1}{\sqrt{N}}m_i\right)(z) = \sum_{i=0}^{N-1} \mathbb{E}(f\overline{m_i})(z)F(m_i)(z).$$

Then looking at the first component (since $r(z) = r(z)$, one can take $w = z$), we get (4.20). $\qquad\square$

Definition 4.7. [DPS14] We denote by L^* the set of all finite words with digits in L, including the empty word. For $l \in L$, let S_l be given as in (4.16) where m_l is replaced by the exponential e_l. If $w = l_1 l_2 \cdots l_n \in L^*$, then by S_w we denote the composition $S_{l_1} S_{l_2} \cdots S_{l_n}$.

Theorem 4.8. [DPS14] *Let* $B \subset \mathbb{R}$, $0 \in B$, $|B| = N$, $R > 1$, *and let* μ_B *be the invariant measure associated to the IFS* $\tau_b(x) = R^{-1}(x+b)$, $b \in B$. *Assume that the IFS has no overlap and that the set* $\frac{1}{R}B$ *has a spectrum* $L \subset \mathbb{R}$, $0 \in L$. *Then the set*

$$\mathcal{E}(L) = \{S_w e_{-c} : c \text{ is an extreme cycle point for } (B, L), w \in L^*\}$$

is an orthonormal basis in $L^2(\mu_B)$. *Some of the vectors in* $\mathcal{E}(L)$ *are repeated, but we count them only once.*

Proof. Let c be an extreme cycle point. Then $|m_B(c)| = 1$. Using the fact that we have equality in the triangle inequality $(1 = |m_B(c)| \leq \frac{1}{N}\sum_{b \in B}|e^{2\pi i bc}| = 1)$, and since $0 \in B$, we get that $e^{2\pi i bc} = 1$ so $bc \in \mathbb{Z}$ for all $b \in B$. Also there exists another extreme cycle point d and $l \in L$ such that $\frac{d+l}{R} = c$. Then we have: $S_l e_{-c}(x) = e^{2\pi i lx}e^{2\pi i(Rx-b)(-c)}$, if $x \in \tau_b(X_B)$. Since $bc \in \mathbb{Z}$ and $R(-c) + l = -d$, we obtain

$$S_l e_{-c} = e_{-d}. \tag{4.21}$$

We use this property to show that the vectors $S_w e_{-c}$, $S_{w'} e_{-c'}$ are either equal or orthogonal for w, w' in L^*, and c, c' extreme cycle points for (B, L).

Using (4.21), we can append some letters at the end of w and w' such that the new words have the same length:

$$S_w e_{-c} = S_{w\alpha} e_{-d}, \quad S_{w'} e_{-c'} = S_{w'\beta} e_{-d'},$$

$$|w\alpha| = |w'\beta| \quad \text{where } d, d' \text{ are cycle points.}$$

Moreover, repeating the letters for the cycle points d and d' as many times as we want, we can assume that α ends in a repetition of the letters associated to d and similarly for β and d'. But, since $|w\alpha| = |w'\beta|$, the Cuntz relations imply that $S_{w\alpha} e_{-d} \perp S_{w'\beta} e_{-d'}$ or $w\alpha = w'\beta$. Assume $|w| \leq |w'|$. Then $\alpha = w''\beta$ for some word w''. Then $S_{w\alpha} e_{-d} \perp S_{w'\beta} e_{-d}$ iff $S_\alpha e_{-d} \perp S_{w''\beta} e_{-d'}$. Also, α consists of repetitions of the digits of the cycle associated to d and similarly for d'. So $S_\alpha e_{-d} = e_{-f}$, $S_{w''\beta} e_{-d'} = e_{-f'}$, and all points d, d', f, f', c, c' belong to the same cycle. So the only case when $S_w e_{-c}$ is not orthogonal to $S_{w'} e_{-c'}$ is when they are equal.

Next, we check that the hypotheses of Theorem 4.1 are satisfied. We let $f(t) = e_{-t} \in L^2(\mu_B)$. To check (i), we just have to see that $e_{-c} \in \cup_{l \in L} S_l \mathcal{E}(L)$. But this follows from (4.21). Requirement (ii) is clear. For (iii), we compute

$$S_l^* e_{-t}(x) = \frac{1}{N} \sum_{b \in B} e^{-2\pi i l \cdot \frac{1}{R}(x+b)} e^{-2\pi i t \cdot \frac{1}{R}(x+b)}$$

$$= e^{-2\pi x \cdot \frac{1}{R}(t+l)} \frac{1}{N} \sum_{b \in B} e^{-2\pi i b(\frac{t+l}{R})}$$

$$= \overline{m_B}\left(\frac{t+l}{R}\right) e_{-\frac{t+l}{R}}(x).$$

So (iii) is satisfied with $\mathfrak{m}_l(t) = \overline{m_B}(\frac{t+l}{R})$, $g_l(t) = \frac{t+l}{R}$.

For (iv), take $c_0 = -c$ for any extreme cycle point (0 is always one). For (v), take h continuous on $\mathbb{R}$, $0 \leq h \leq 1$, $h(c) = 1$ for all c with $e_{-c} \in \overline{\text{span}}\mathcal{E}(L)$, and

$$h(t) = \sum_{l \in L} \left| m_B\left(\frac{t+l}{R}\right)\right|^2 h\left(\frac{t+L}{R}\right) := Rh(t).$$

In particular, we have $h(c) = 1$ for every extreme cycle point c. Assume $h \not\equiv 1$. First we will restrict our attention to $t \in I := [a, b]$ with $a \leq \frac{\min L}{R-1}$, $b \geq \frac{\max L}{R-1}$, and note that $g_l(I) \subset I$ for all $l \in L$. Let $m = \min_{t \in I} h(t)$. Then let $h' = h - m$, assume $m < 1$. Then $Rh'(t) = h'(t)$ for all $t \in \mathbb{R}$,

h' has a zero in I, and $h \geq 0$ on I, $h'(z_0) = 0$. But this implies that $|m_B(g_l(z_0))|^2 h'(g_l(z_0)) = 0$ for all $l \in L$. Since $\sum_{l \in L} |m_B(g_l(z_0))|^2 = 1$, it follows that for one of the $l_0 \in L$ we have $h'(g_{l_0}(z_0)) = 0$. By induction, we can find $z_n = g_{l_{n-1}} \cdots g_{l_0} z_0$ such that $h'(z_n) = 0$. We prove that z_0 is a cycle point. Suppose not. Since m_B has finitely many zeros, for n large enough $g_{\alpha_k} \cdots g_{\alpha_1} z_n$ is not a zero for m_B, for any choice of digits $\alpha_1, \ldots, \alpha_k$ in L. But then, by using the same argument as above we get that $h'(g_{\alpha_k} \cdots g_{\alpha_1} z_n) = 0$ for any $\alpha_1, \ldots, \alpha_k \in L$. The points $\{g_{\alpha_k} \cdots g_{\alpha_1} z_n : \alpha_1, \ldots, \alpha_k \in L, k \in \mathbb{N}\}$ are dense in the attractor X_L of the IFS $\{g_l\}_{l \in L}$, thus h' is constant 0 on X_L. But the extreme cycle points c are in X_L and since $h(c) = 1$, we have $0 = h'(c) = 1 - m$, so $m = 1$. Thus, $h = 1$ on I. Since we can let $a \to -\infty$, and $b \to \infty$, we obtain that $h \equiv 1$. $\qquad\square$

Remark 4.9. [DPS14] The functions in $\mathcal{E}(L)$ are piecewise exponential. The formula for $S_{l_1 \cdots l_n} e_{-c}$ is

$$S_{l_1 \cdots l_n} e_{-c}(x) = e^{\alpha(b,l,c)} \ldots e_{l_1 + Rl_2 + \cdots + R^{n-1} l_{n-1} + R^n(-c)}(x), \qquad (4.22)$$

where $\alpha(b, l, c) = -[b_1 l_2 + (Rb_1 + b_2) l_3 + \cdots + (R^{n-2} b_1 + \cdots + b_{n-1}) l_n] + (R^{n-1} b_1 + \cdots + b_n) \cdots c$ if $x \in \tau_{b_1} \cdots \tau_{b_n} X_B$. We have

$$S_{l_1} \cdots S_{l_n} e_{-c}(x) = e_{l_1}(x) e_{l_2}(rx) \cdots e_{l_n}(r^{n-1} x) e_c(r^n x)$$

If $x \in \tau_{b_1} \cdots \tau_{b_n} X_B$, then $rx \in \tau_{b_2} \cdots \tau_{b_n} X_B$, $r^{n-1} x \in \tau_{b_n} X_B$. So

$$rx = Rx - b_1$$

$$r^2 x = Rrx - b_2 = R^2 x - Rb_1 - b_2$$

$$\vdots$$

$$r^{n-1} x = R^{n-1} x - R^{n-2} b_1 - \cdots - Rb_{n-2} - b_{n-1}$$

$$r^n x = R^n x - R^{n-1} b_1 - R^{n-2} b_2 - \cdots - Rb_{n-1} - b_n.$$

The rest follows from a direct computation.

Corollary 4.10. [DPS14] *In the hypothesis of Theorem 4.1, if in addition $B, L \subset \mathbb{Z}$ and $R \in \mathbb{Z}$, then there exists a set Λ such that $\{e_\lambda : \lambda \in \Lambda\}$ is an orthonormal basis for $L^2(\mu_B)$.*

Proof. If everything is an integer, then it follows from Remark 4.9 that $S_w e_{-c}$ is an exponential function for all w and extreme cycle points c. Note that, as in the proof of Theorem 4.1, $bc \in \mathbb{Z}$ for all $b \in B$. $\qquad\square$

Example 4.11. [DPS14] We consider the IFS that generates the middle third Cantor set: $R = 3$, $B = \{0,2\}$. The set $\frac{1}{3}\{0,2\}$ has spectrum $L = \{0,3/4\}$. We look for the extreme cycle points for (B,L).

We need $|m_B(-c)| = 1$ so $|\frac{1+e^{2\pi i 2c}}{2}| = 1$, therefore $c \in \frac{1}{2}\mathbb{Z}$. Also c has to be a cycle for the IFS $g_0(x) = x/3$, $g_{3/4}(x) = \frac{x+3/4}{3}$ so $0 \le c \le \frac{3/4}{3-1} = 3/8$. Thus, the only extreme cycle is $\{0\}$. By Theorem 4.1, $\mathcal{E} = \{S_w 1 : w \in \{0,3/4\}^*\}$ is an orthonormal basis for $L^2(\mu_B)$. Note also that the numbers $e^{2\pi i \alpha(b,l,c)}$ in formula (4.22) are ± 1 because $2\pi i B \cdot L \subset \pi i \mathbb{Z}$.

4.7.1.2 *Walsh bases*

In the following, we will focus on the unit interval, which can be regarded as the attractor of a simple IFS and we use step functions for the QMF basis to generate Walsh-type bases for $L^2[0,1]$ [DPS14].

Example 4.12. [DPS14] The interval $[0,1]$ is the attractor of the IFS $\tau_0 x = \frac{x}{2}$, $\tau_1 x = \frac{x+1}{2}$, and the invariant measure is the Lebesgue measure on $[0,1]$. The map r defined in Example 4.3 is $rx = 2x \bmod 1$. Let $m_0 = 1$, $m_1 = \chi_{[0,1/2)} - \chi_{[1/2,1)}$. It is easy to see that $\{m_0, m_1\}$ is a QMF basis. Therefore, S_0, S_1 defined as in Proposition 4.5 form a representation of the Cuntz algebra $\mathcal{O}_2$.

Proposition 4.13. [DPS14] *The set $\mathcal{E} := \{S_w 1 : w \in \{0,1\}^*\}$ is an orthonormal basis for $L^2[0,1]$, the Walsh basis.*

Proof. We check the conditions in Theorem 4.1. To see that (i) holds, note that $S_0 1 = 1$. Define $f(t) = e_t$, $t \in \mathbb{R}$. (ii) is clear. For (iii), we compute

$$S_1^* e_t(x) = \frac{1}{2}(e^{2\pi i t \cdot x/2} + e^{2\pi i t \cdot (x+1)/2}) = e^{2\pi i t \cdot x/2} \frac{1}{2}(1 + e^{2\pi i t/2})$$

$$S_1^* e_t(x) = \frac{1}{2}(e^{2\pi i t \cdot x/2} - e^{2\pi i t \cdot (x+1)/2}) = e^{2\pi i t \cdot x/2} \frac{1}{2}(1 - e^{2\pi i t/2}).$$

Thus, (iii) holds with $m_0(t) = \frac{1}{2}(1 + e^{2\pi i t/2})$, $m_1(t) = \frac{1}{2}(1 - e^{2\pi i t/2})$, $g_0(t) = g_1(t) = \frac{t}{2}$. Since $e_0 = 1$, it follows that (iv) holds.

For (v), take h continuous on $\mathbb{R}$, $0 \le h \le 1$, $h(c) = 1$ for all $c \in \mathbb{R}$ with $e_t \in \overline{\text{span}}\mathcal{E}$, in particular $h(0) = 1$ and

$$h(t) = \left|\frac{1}{2}(1 + e^{2\pi it/2})\right|^2 h(t/2) + \left|\frac{1}{2}(1 - e^{2\pi it/2})\right|^2 h(t/2) = h(t/2).$$

Then $h(t) = h(t/2^n)$ for all $t \in \mathbb{R}$, $n \in \mathbb{N}$. Letting $n \to \infty$ and using the continuity of h, we get $h(t) = h(0) = 1$ for all $t \in \mathbb{R}$. Since all conditions hold, we get that $\mathcal{E}$ is an orthonormal basis. That $\mathcal{E}$ is actually the Walsh basis follows from the following calculations: for $|w| = n$ in $\{0,1\}^*$, let $n = \sum_i x_i 2^i$ be the base 2 expansion of n. Because $S_0 f = f \circ r$, $S_1 f = m_1 f \circ r$, and $m_0 \equiv 1$, we obtain the following decomposition:

$$S_w 1(x) = m_1(r^{i_1}x) \cdot m_1(r^{i_2}x) \cdots m_1(r^{i_k}x),$$

where $i_1, i_2, \ldots, i_k$ correspond to those i with $x_i = 1$.

Also $m_1(r^i x) = m_1(2^i x \bmod i)$ are the Rademacher functions and thus we obtain the Walsh basis (see, e.g., [SWS90]). $\qquad\square$

The Walsh bases can be easily generalized by replacing the matrix

$$\frac{1}{\sqrt{2}}\begin{pmatrix} 1 & 1 \\ 1 & -1 \end{pmatrix},$$

which appears in the definition of the filters m_0, m_1, with an arbitrary unitary matrix A with constant first row and by changing the scale from 2 to N.

Theorem 4.14. [DPS14] *Let $N \in \mathbb{N}$, $N \ge 2$. Let $A = [a_{ij}]$ be an $N \times N$ unitary matrix whose first row is constant $\frac{1}{\sqrt{N}}$. Consider the IFS $\tau_j x = \frac{x+j}{N}$, $x \in \mathbb{R}$, $j = 0, \ldots, N-1$ with the attractor $[0,1]$ and invariant measure the Lebesgue measure on $[0,1]$. Define*

$$m_i(x) = \sqrt{N} \sum_{j=0}^{N-1} a_{ij} \chi_{[j/N,(j+1)/N]}(x).$$

Then $\{m_i\}_{i=0}^{N-1}$ is a QMF basis. Consider the associated representation of the Cuntz algebra $\mathcal{O}_N$. Then the set $\mathcal{E} := \{S_w 1 : w \in \{0, ..., N-1\}^*\}$ is an orthonormal basis for $L^2[0,1]$.

Proof. We check the conditions in Theorem 4.1. Let $f(t) = e_t$, $t \in \mathbb{R}$. To check (i), note that $S_0 1 \equiv 1$. (ii) is clear. For (iii), we compute:

$$S_k^* e_t = \frac{1}{N} \sum_{j=0}^{N-1} \overline{m_k}(\tau_j x) e_t(\tau_j x)$$

$$= \frac{1}{\sqrt{N}} \sum_{j=0}^{N-1} \overline{a_{kj}} e^{2\pi i t \cdot (x+j)/N}$$

$$= e^{2\pi i t \cdot x/N} \frac{1}{\sqrt{N}} \sum_{j=0}^{N-1} \overline{a_{kj}} e^{2\pi i t \cdot j/N}.$$

So (iii) is true with $\mathsf{m}_k(t) = \frac{1}{\sqrt{N}} \sum_{j=0}^{N-1} \overline{a_{kj}} e^{2\pi i t \cdot j/N}$ and $g_k(t) = \frac{t}{N}$. (iv) is true with $c_0 = 0$. For (v), take $h \in \mathcal{C}(\mathbb{R})$, $0 \le h \le 1$, $h(c) = 1$ for all $c \in \mathbb{R}$ with $e_c \in \overline{\operatorname{span}\mathcal{E}}$ (in particular $h(0) = 1$), and

$$h(t) = \sum_{k=0}^{N-1} |\mathsf{m}_k(t)|^2 h(t/N)$$

$$= h(t/N) \sum_{k=0}^{N-1} \frac{1}{N} |\sum_{j=0}^{N-1} a_{kj} e^{-2\pi i t \cdot j/N}|^2$$

$$= h(t/N) \cdot \frac{1}{N} \|Av\|^2.$$

where $v = (e^{-2\pi i t \cdot j/N})_{j=0}^{N-1}$. Since A is unitary, $\|Av\|^2 = \|v\|^2 = N$. Then $h(t) = h(t/N^n)$. Letting $n \to \infty$ and using the continuity of h, we obtain that $h(t) = 1$ for all $t \in \mathbb{R}$. Thus, Theorem 4.1 implies that $\mathcal{E}$ is an orthonormal basis. $\square$

Remark 4.15. [DPS14] We can read the constants that appear in the step function $S_w 1$ from the tensor of A with itself n times, where n is the length of the word w.

Let A be an $N \times N$ matrix, B, an $M \times M$ matrix. Then $A \otimes B$ has the following entries:

$$(A \otimes B)_{i_1 + M i_2, j_1 + M j_2} = a_{i_1 j_1} b_{i_2 j_2},$$
$$i_1, j_1 = 0, \ldots, N-1, \ i_2, j_2 = 0, \ldots, M-1$$

$$A \otimes B = \begin{pmatrix} Ab_{0,0} & Ab_{0,1} & \cdots & Ab_{0,M-1} \\ Ab_{1,0} & Ab_{1,1} & \cdots & Ab_{1,M-1} \\ \vdots & \vdots & \ddots & \vdots \\ Ab_{M-1,0} & Ab_{M-1,1} & \cdots & Ab_{M-1,M-1} \end{pmatrix}.$$

The matrix $A^{\otimes n}$ is obtained by induction, tensoring to the left: $A^{\otimes n} = A \otimes A^{\otimes(n-1)}$.

Thus, $A \otimes A \otimes A \otimes \cdots \otimes A$, n times, has entries

$$A^{\otimes n}_{i_0+Ni_1+N^2 i_2+\cdots+N^{n-1}i_{n-1},\,j_0+Nj_1+\cdots+N^{n-1}j_{n-1}} = a_{i_0 j_0} a_{i_1 j_1} \cdots a_{i_{n-1} j_{n-1}}.$$

Now compute for $i_0,\ldots,i_{n-1} \in \{0,\ldots,N-1\}$:

$$S_{i_0 \cdots i_{n-1}} 1(x) = m_{i_0}(x) m_{i_1}(rx) \cdots m_{i_{n-1}}(r^{n-1}x).$$

Suppose $x \in [\frac{k}{N^n}, \frac{k+1}{N^n})$, $0 \le k < N^n$ and $k = N^{n-1}j_0 + N^{n-2}j_1 + \cdots + N j_{n-2} + j_{n-1}$, where $0 \le j_0,\ldots,j_{n-1} < N$.

Then $x \in [\frac{j_0}{N}, \frac{j_0+1}{N})$, $rx = (Nx) \bmod 1 \in [\frac{j_1}{N}, \frac{j_1+1}{N}),\ldots, r^{n-1}x = (N^{n-1}x) \bmod 1 \in [\frac{j_{n-1}}{N}, \frac{j_{n-1}+1}{N})$, so $m_{i_0}(x) = \sqrt{N} a_{i_0 j_0}$, $m_{i_1}(rx) = \sqrt{N} a_{i_1 j_1}, \ldots, m_{i_{n-1}}(r^{n-1}x) = \sqrt{N} a_{i_{n-1} j_{n-1}}$ hence,

$$S_{i_0 \cdots i_{n-1}} 1(x) = \sqrt{N^n} a_{i_0 j_0} \cdots a_{i_{n-1} j_{n-1}}$$

$$= \sqrt{N^n} A^{\otimes n}_{i_0+Ni_1+N^2 i_2+\cdots+N^{n-1}i_{n-1},\,j_0+Nj_1+\cdots+N^{n-1}j_{n-1}}.$$

Example 4.16. [DPS14] The pictures in Figure 4.7 show the Walsh functions that correspond to the scale $N = 4$ and the matrix

$$A = \begin{pmatrix} \frac{1}{2} & \frac{1}{2} & \frac{1}{2} & \frac{1}{2} \\ \frac{\sqrt{2}}{2} & -\frac{\sqrt{2}}{2} & 0 & 0 \\ 0 & 0 & \frac{\sqrt{2}}{2} & -\frac{\sqrt{2}}{2} \\ \frac{1}{2} & \frac{1}{2} & -\frac{1}{2} & -\frac{1}{2} \end{pmatrix}$$

for the words of length 2, indicated at the top.

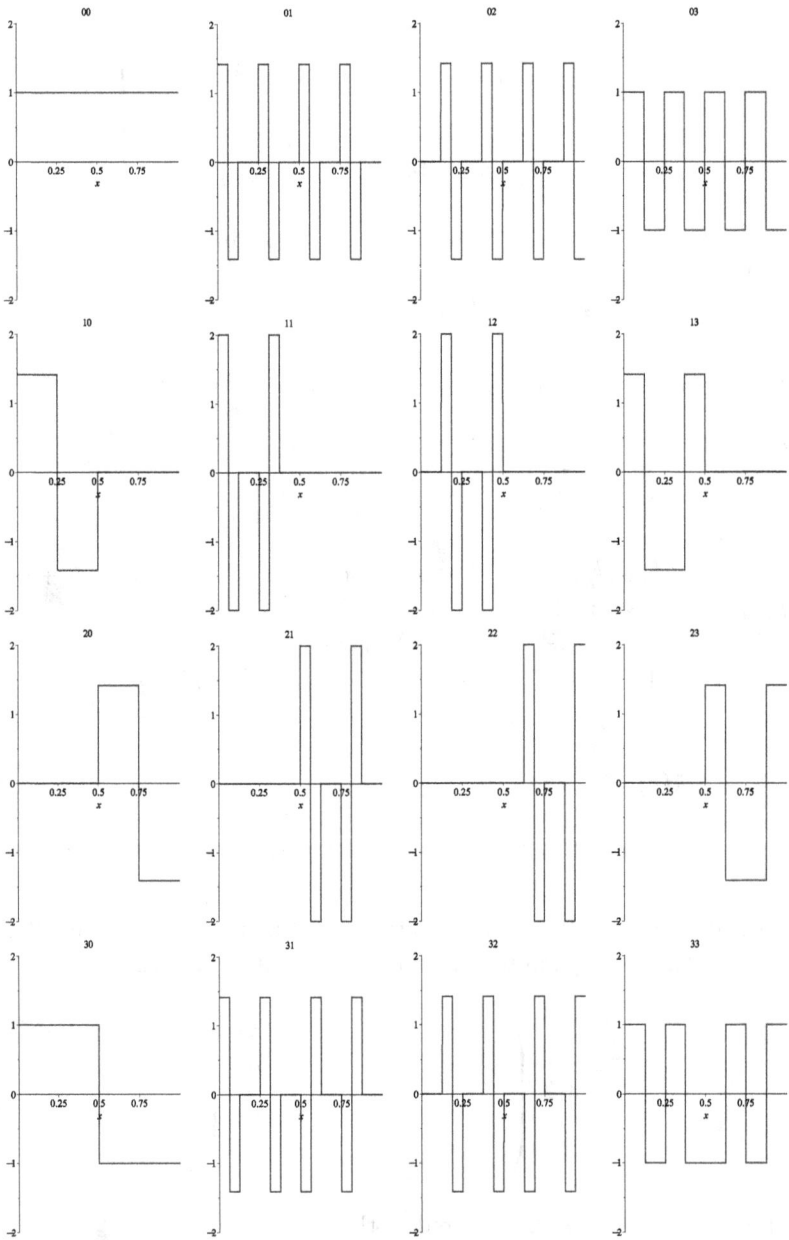

Fig. 4.7 Walsh functions $S_w 1$ for words w of length 2 [DPS14].

4.8 List of Names and Discoveries

Many of the main discoveries summarized below are now lore. Also, see [Dau92, Bri10a, Bri10b, Bri13a, Bri13b, Sha93a, Sha57, dLMSS56, HKMMT89, GM85, JMR01, Mey00a, Coh03a, CDDD03, Coh03b, CCM16, AMWW16, CD04, DDLY00, Don00].

1807 Jean Baptiste Joseph Fourier mathematics, physics (heat conduction)	Expressed functions as sums of sine and cosine waves of frequencies in arithmetic progression (now called Fourier series).
1909 Alfred Haar mathematics	Discovered, while a student of David Hilbert, an orthonormal basis consisting of step functions, applicable both to functions on an interval, and functions on the whole real line. While it was not realized at the time, Haar's construction was a precursor of what is now known as the Mallat subdivision, and multiresolution method, as well as the subdivision wavelet algorithms.
1946 Denes Gabor (Nobel Prize): physics (optics, holography)	Discovered basis expansions for what might now be called time-frequency wavelets, as opposed to time-scale wavelets.
1948 Claude Elwood Shannon mathematics, engineering (information theory)	A rigorous formula used by the phone company for sampling speech signals. Quantizing information, entropy, founder of what is now called the mathematical theory of communication.
1976 Claude Garland, Daniel Esteban (both) signal processing	Discovered subband coding of digital transmission of speech signals over the telephone.

1981
Jean Morlet
petroleum engineer

Suggested the term "ondelettes." J.M. decomposed reflected seismic signals into sums of "wavelets (Fr.: ondelettes) of constant shape," i.e., a decomposition of signals into wavelet shapes, selected from a library of such shapes (now called wavelet series). Received somewhat late recognition for his work. Due to contributions by A. Grossman and Y. Meyer, Morlet's discoveries have now come to play a central role in the theory.

1985
Yves Meyer
mathematics,
applications

Mentor for A. Cohen, S. Mallat, and the other wavelet pioneers, Y.M. discovered infinitely often differentiable wavelets.

1989
Albert Cohen
mathematics (ortho-
gonality relations),
numerical analysis

Discovered the use of wavelet filters in the analysis of wavelets — the so-called Cohen condition for orthogonality.

1986
Stéphane Mallat
mathematics, signal
and image processing

Discovered what is now known as the subdivision, and multiresolution method, as well as the subdivision wavelet algorithms. This allowed the effective use of operators in the Hilbert space $L^2(\mathbb{R})$, and of the parallel computational use of recursive matrix algorithms.

1987
Ingrid Daubechies
mathematics, physics,
and communications

Discovered differentiable wavelets, with the number of derivatives roughly half the length of the support interval. Further found polynomial algorithms for their construction (with coauthor Jeff Lagarias; joint spectral radius formulas).

1991	Discovered the use of a transfer operator in
Wayne Lawton	the analysis of wavelets: orthogonality and
mathematics	smoothness.
(the wavelet	
transfer operator)	

1991
Wayne Lawton
mathematics
(the wavelet
transfer operator)

Discovered the use of a transfer operator in
the analysis of wavelets: orthogonality and
smoothness.

1992
The FBI
using wavelet algo-
rithms in digitizing
and compressing
fingerprints

C. Brislawn and his group at Los Alamos
created the theory and the codes which allowed
the compression of the enormous FBI fingerprint file, creating A/D, a new database of
fingerprints.

2000
The International
Standards
Organization

A wavelet-based picture compression standard,
called JPEG 2000, for digital encoding of
images.

1994
David Donoho
statistics,
mathematics

Pioneered the use of wavelet bases and tools
from statistics to "denoise" images and signals.

4.9 History

While wavelets as they have appeared in the mathematics literatures
(e.g., [Dau92, Bri10a, Bri10b, Bri13a, Bri13b, Sha93a, Sha57, dLMSS56,
HKMMT89, GM85, JMR01, Mey00a, Coh03a, CDDD03, Coh03b, CCM16,
AMWW16, CD04, DDLY00, Don00]) for a long time, starting with Haar in
1909, involve function spaces, the connections to a host of discrete problems
from engineering are more subtle. Moreover, the deeper connections
between the discrete algorithms and the function spaces of mathematical
analysis are of a more recent vintage, see, e.g., [SN96] and [Jor06a].

Here, we begin with the function spaces. This part of wavelet theory refers to continuous wavelet transforms (details in what follow).

It dominated the wavelet literature in the 1980s, and is beautifully treated in the first four chapters in [Dau92] and in [Dau93]. The word "continuous" refers to the continuum of the real line $\mathbb{R}$. Here, we consider spaces of functions in one or more real dimensions, i.e., functions on the line $\mathbb{R}$ (signals), the plane $\mathbb{R}^2$ (images), or in higher dimensions $\mathbb{R}^d$, functions of d real variables.

4.10 Literature

As evidenced by a simple Google check, the mathematical wavelet literature is gigantic in size, and the manifold applications spread over a vast number of engineering journals. While we cannot do justice to this voluminous literature, we instead offer a collection of the classics [HW06] edited recently by C. Heil *et al.*

Chapter 5

Entropy Encoding, Hilbert Space, and Karhunen–Loève Transforms

In this chapter, we discuss entropy encoding schemes in wavelet image decomposition step using Karhunen–Loève Transforms, some of which are "borrowed" from [JS07, Son07]. Entropy encoding is mentioned in Chapter 3, and is performed after the wavelet image decomposition step. Also, we discuss Karhunen–Loève Transforms, which are also called principal component analysis (PCA) by engineers in theory and algorithms. We show specific examples to illustrate PCA in Section 5.9.

5.1 Introduction

Historically, the Karhunen–Loève transform arose as a tool from the interface of probability theory and information theory; see details with references inside the book. It has served as a powerful tool in a variety of applications; starting with the problem of separating variables in stochastic processes, say X_t; processes that arise from statistical noise, e.g., from fractional Brownian motion. Since the initial inception in mathematical statistics, the operator algebraic contents of the arguments have crystallized as follows: starting from the process X_t, for simplicity assume zero mean, i.e., $E(X_t) = 0$; create a correlation matrix $C(s,t) = E(X_s X_t)$. (Strictly speaking, it is not a matrix, but rather an integral kernel. Nonetheless, the matrix terminology has stuck.) The next key analytic step in the Karhunen–Loève method is to then apply the Spectral Theorem from operator theory to a corresponding self-adjoint operator, or to some operator naturally associated with C: hence the name, the Karhunen–Loève Decomposition (KLC). In favorable cases (discrete spectrum), an orthogonal family of functions $(f_n(t))$ in the time variable arise, and a corresponding family of eigenvalues. We take them to be normalized in a suitably chosen square-norm. By integrating the basis functions $f_n(t)$ against X_t, we get a sequence of random variables Y_n. It was the insight of Karhunen–Loève [Loe55] to give general conditions for when this sequence of random variables is independent, and to show that if the initial random process X_t is Gaussian, then so are the random variables Y_n. (See also Example 5.1 that follows.)

In the 1940s, Kari Karhunen ([Kar46], [Kar52]) pioneered the use of spectral theoretic methods in the analysis of time series, and more generally in stochastic processes. It was followed up by papers and books by Michel Loève in the 1950s [Loe55], and in 1965 by R.B. Ash [Ash90]. (Note that this theory precedes the surge in the interest in wavelet bases!)

As we outline in what follows, all the settings place rather stronger assumptions. We argue how more modern applications dictate more general theorems, which we prove in our book. A modern tool from operator theory and signal processing, which we will use, is the notion of *frames* in Hilbert space. More precisely, frames are redundant "bases" in Hilbert space. They are called framed, but intuitively should be thought of as generalized bases. The reason for this, as we show, is that they offer an explicit choice of a (non-orthogonal) expansion of vectors in the Hilbert space under consideration.

In our book, we rely on the classical literature (see, e.g., [Ash90]), and we accomplish three things: (i) We extend the original Karhunen–Loève idea to the case of continuous spectrum; (ii) we give frame theoretic uses of the Karhunen–Loève idea which arise in various wavelet contexts and which go beyond the initial uses of Karhunen–Loève; and finally (iii) we give applications.

These applications in our case come from image analysis; specifically from the problem of statistical recognition and detection; e.g., to nonlinear variance, e.g., due to illumination effects. Then the Karhunen–Loève Decomposition (KLD), also known as Principal Component Analysis (PCA), applies to the intensity images. This is traditional in statistical signal detection and in estimation theory. Adaptations to compression and recognition are of a more recent vintage. In brief outline, each intensity image is converted into vector form. (This is the simplest case of a purely intensity-based coding of the image, and it is not necessarily ideal for the application of KL-decompositions.)

The ensemble of vectors used in a particular conversion of images is assumed to have a multivariate Gaussian distribution since human faces form a dense cluster in image space. The PCA method generates small set of basis vectors forming subspaces whose linear combinations offer better (or perhaps ideal) approximations to the original vectors in the ensemble. In facial recognition, the new bases are said to span intra-face and inter-face variations, permitting Euclidean distance measurements to exclusively pick up changes in, e.g., identity and expression.

Our presentation will start with various operator theoretic tools, including frame representations in Hilbert space. We have included more details and more explanations than is customary in more narrowly focused papers, as we wish to cover the union of four overlapping fields of specialization: operator theory, information theory, wavelets, and physics applications. The novice might find the following references helpful: [BCW90, BR91].

While entropy encoding is popular in engineering, [SCE01], [Use01], [DVDD98], the choices made in signal processing are often more by trial and error than by theory. Reviewing the literature, we found that the mathematical foundation of the current use of entropy in encoding deserves closer attention.

In this book, we take advantage of the fact that Hilbert space and operator theory form the common language of both quantum mechanics and of signal/image processing. Recall first that in quantum mechanics, (pure) states as mathematical entities "are" one-dimensional subspaces in complex Hilbert space $\mathcal{H}$, so we may represent them by vectors of norm one. Observables "are" self-adjoint operators in $\mathcal{H}$, and the measurement problem entails von Neumann's spectral theorem applied to the operators.

In signal processing, time-series or matrices of pixel numbers may similarly be realized by vectors in Hilbert space $\mathcal{H}$. The probability distribution of quantum mechanical observables (state space $\mathcal{H}$) may be represented by choices of orthonormal bases (ONBs) in $\mathcal{H}$ in the usual way (see, e.g., [Jor06a]). In signal/image processing, because of aliasing, it is practical to generalize the notion of ONB, and this takes the form of what is called "a system of frame vectors"; see [Chr03].

But even von Neumann's measurement problem, viewing experimental data as part of a bigger environment (see, e.g., [DS97], [Wan04], [Eld03]) leads to basis notions more general than ONBs. They are commonly known as Positive Operator Valued Measures (POVMs), and in the present book we examine the common ground between the two seemingly different uses of operator theory in the separate applications. To make the book presentable to two audiences, we have included a few more details than is customary in pure math papers.

We show that parallel problems in quantum mechanics and in signal processing entail the choice of "good" orthonormal bases (ONBs). One particular such ONB goes under the name "the Karhunen–Loève basis." We will show that it is optimal in three ways, and we will outline a number of applications.

The problem addressed in this book is motivated by consideration of the optimal choices of bases for certain analogue-to-digital (A-to-D) problems we encountered in the use of wavelet bases in image processing (see [GWE04], [SCE01], [Use01], [Wal99b]); but certain of our considerations have an operator theoretic flavor, which we wish to isolate, as it seems to be of independent interest.

There are several reasons why we take this approach. First, our Hilbert space results seem to be of general interest outside the particular applied context where we encountered them. And second, we feel that our more abstract results might inspire workers in operator theory and approximation theory.

5.1.1 *Digital image compression*

In digital image compression, after quantization (see Figure 5.1) entropy encoding is performed on a particular image for more efficient-less storage memory-storage. When an image is to be stored, we need either 8 bits or 16 bits to store a pixel. With efficient entropy encoding, we can use a smaller number of bits to represent a pixel in an image, resulting in less memory used to store or even transmit an image. Thus, the Karhunen–Loève theorem enables us to pick the best basis thus to minimize the entropy and error, to better represent an image for optimal storage or transmission. Here, *optimal* means it uses least memory space to represent the data.

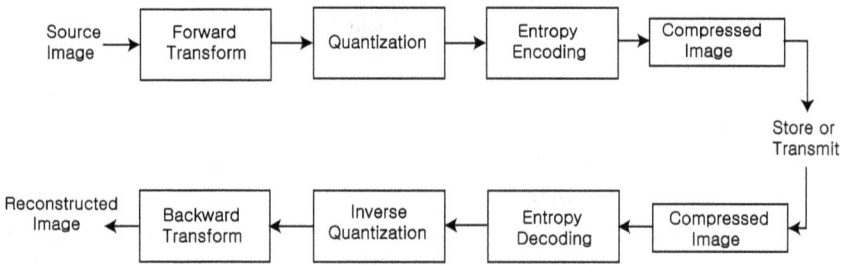

Fig. 5.1 Outline of the wavelet image compression process [SCE01].

That is, instead of using 16 bits, it uses 11 bits. So, the best basis found would allow us to better represent the digital image with less storage memory.

Recall, Figure 2.9.

In our next section, we give the general context and definitions from operators in Hilbert space, which we shall need: We discuss the particular orthonormal bases (ONBs) and frames which we use, and we recall the operator theoretic context of the Karhunen–Loève theorem [Ash90]. In approximation problems involving a stochastic component (e.g., noise removal in time-series or data resulting from image processing), one typically ends up with correlation kernels; in some cases, frame kernels; see details in Section D.3. In some cases, they arise from systems of vectors in Hilbert space, which form frames (see Definition 5.5). In some cases, parts of the frame vectors fuse (fusion frames) onto closed subspaces, and we will be working with the corresponding family of (orthogonal) projections. Either way, we arrive at a family of self-adjoint positive semidefinite operators in Hilbert space. The particular Hilbert space depends on the application at hand. While the Spectral Theorem does allow us to diagonalize these operators, the direct application of the Spectral Theorem may lead to continuous spectrum, which is not directly useful in computations, or it may not be computable by recursive algorithms. As a result, we introduce in Section 5.6 a weighting of the operator to be analyzed.

The questions we address are optimality of approximation in a variety of ONBs, and the choice of the "best" ONB. Here, "best" is given two precise meanings: (1) In the computation of a sequence of approximations to the frame vectors, the error terms must be smallest possible; and similarly, (2) we wish to minimize the corresponding sequence of entropy numbers (referring to von Neumann's entropy). In two theorems, we make precise an operator theoretic Karhunen–Loève basis, which we show is optimal both in regards to criteria (1) and (2). But before we prove our theorems, we give the two problems an operator theoretic formulation; and in fact our theorems are stated in this operator theoretic context.

In Section 5.6, we introduce the weighting, and we address a third optimality criteria; that of optimal weights: Among all the choices of weights (taking the form of certain discrete probability distributions) turning the initially given operator into trace-class, the problem is then to select the particular weights which are optimal in a sense which we define precisely.

5.2 General Background

5.2.1 *From data to Hilbert space*

In computing probabilities and entropy, Hilbert space serves as a helpful tool. As an example, take a unit vector f in some fixed Hilbert space $\mathcal{H}$, and an orthonormal basis (ONB) ψ_i with i running over an index set I. With this, we now introduce two families of probability measures, one family $P_f(\cdot)$ indexed by $f \in \mathcal{H}$, and a second family P_T indexed by a class of operators $T : \mathcal{H} \to \mathcal{H}$.

5.2.1.1 *The measures P_f*

Define

$$P_f(A) = \sum_{i \in A} |\langle \psi_i | f \rangle|^2, \tag{5.1}$$

where $A \subset I$, and where $\langle \cdot | \cdot \rangle$ denotes the inner product. Following physics convention, we make our inner product linear in the second variable. That will also let us make use of Dirac's convenient notation for rank-one operators, see Eq. (5.41).

Note then that $P_f(A)$ is a probability measure on the finite subsets A of I. To begin with, we make the restriction to finite subsets. This is merely for later use in recursive systems, see, e.g., Eq. (5.2). In diverse contexts, extensions from finite to infinite is are then done by means of Kolmogorov's consistency principle [Kol77].

By introducing a weighting, we show that this assignment also works for more general vector configurations $\mathcal{C}$ than ONBs. Vectors in $\mathcal{C}$ may represent signals or image fragments/blocks. Correlations would then be measured as inner products $\langle u | v \rangle$ with u and v representing different image pixels. Or in the case of signals, u and v might represent different frequency subbands.

5.2.1.2 *The measures P_T*

A second more general family of probability measures arising in the context of Hilbert space is called determinantal measures. Specifically, consider bitstreams as points in an infinite Cartesian product $\Omega = \prod_{i\in\mathbb{N}}\{0,1\}$. Cylinder sets in Ω are indexed by finite subsets $A \subset \mathbb{N}$,

$$C_A = \{(\omega_1, \omega_2, ...)|\omega_i = 1 \quad \text{for } i \in A\}.$$

If T is an operator in $l^2(\mathbb{N})$ such that $0 \le \langle u|Tu \rangle \le \|u\|^2$ for all $u \in l^2$, then set

$$P_T(C_A) = \det(T(i,j))_{i,j\in A}, \tag{5.2}$$

where $T(i,j)$ is the matrix representation of T computed in some ONB in l^2. Using general principles [Kol77, Jor06a], it can be checked that $P_T(C_A)$ is independent of the choice of ONB.

To verify that $P_T(\cdot)$ extends to a probability measure defined on the sigma-algebra generated by C_As, see, e.g., [Jor06a], Ch7. The argument is based on Kolmogorov's consistency principle, see [Kol77]

Frames (Definition 5.6) are popular in analyzing signals and images. This fact raises questions of comparing two approximations: one using a frame, the other using an ONB. However, there are several possible choices of ONBs. An especially natural choice of an ONB would be one that diagonalizes the matrix $(\langle f_i|f_j \rangle)$ where (f_i) is a frame. We call such a choice of ONB Karhunen–Loève (K-L) expansion. Section 5.3 deals with a continuous version of this matrix problem. The justification for why diagonalization occurs and works also when the frame (f_i) is infinite is based on the Spectral Theorem. For the details regarding this, see the following proof of Theorem 5.3.

In symbols, we designate the K-L ONB associated to the frame (f_i) as (ϕ_i). In computations, we must rely on finite sums, and we are interested in estimating the errors when different approximations are used, and where summations are truncated. Our main results make precise how the K-L ONB yields better approximations, smaller entropy, and better synthesis. Even more, we show that infimum calculations yield minimum numbers attained at the K-L ONB expansions. We emphasize that the ONB depends on the frame chosen, and this point will be discussed in detail later.

If larger systems are subdivided, the smaller parts may be represented by projections P_i, and the $i - j$ correlations by the operators $P_i P_j$. The entire family (P_i) is to be treated as a fusion frame [CK04, CKL08]. Fusion frames are defined in Definition 5.15 that follows. Frames themselves are generalized bases with redundancy, e.g., occurring in signal processing involving multiplexing. The fusion frames allow decompositions with closed subspaces as opposed to individual vectors. They allow decompositions of signal/image processing tasks with degrees of homogeneity.

5.3 The Karhunen–Loève Transform

In general, one refers to a *Karhunen–Loève transform* as an expansion in Hilbert space with respect to an ONB resulting from an application of the Spectral Theorem.

Consider an ensemble of a large number N of similar objects, of which Nw^α, $\alpha = 1, 2, ..., \nu$ where the relative frequency w^α satisfies the probability axioms:

$$w^\alpha \geq 0, \quad \sum_{\alpha=1}^{\nu} w^\alpha = 1.$$

Assume that each type specified by a value of the index α is represented by $f^\alpha(\xi)$ in a real domain $[a, b]$, which we normalize by

$$\int_a^b |f^\alpha(\xi)|^2 d\xi = 1.$$

Let $\{\psi_i(\xi)\}$, $i = 1, 2, ...$, be a complete set of orthonormal base functions defined on $[a, b]$. Then any function $f^\alpha(\xi)$ can be expanded as

$$f^\alpha(\xi) = \sum_{i=1}^{\infty} x_i^{(\alpha)} \psi_i(\xi) \tag{5.3}$$

with

$$x_i^\alpha = \int_a^b \psi_i^*(\xi) f^\alpha(\xi) d\xi. \tag{5.4}$$

Here, x_i^α is the component of f^α in ψ_i coordinate system. With the normalization of f^α, we have

$$\sum_{i=1}^{\infty} |x_i^\alpha|^2 = 1. \tag{5.5}$$

Then substituting (5.4) in (5.3) gives

$$f^\alpha(\xi) = \int_a^b f^\alpha(\xi) [\sum_{i=1}^{\infty} \psi_i^*(\xi) \psi_i(\xi)] d\xi = \sum_{i=1}^{\infty} \langle \psi_i(\xi) | f^\alpha \rangle \psi_i \tag{5.6}$$

by definition of ONB.

Let $\mathcal{H} = L^2(a, b)$. $\psi_i : \mathcal{H} \to l^2(\mathbb{Z})$, and $U : l^2(\mathbb{Z}) \to l^2(\mathbb{Z})$ where U is a unitary operator [JS07, Son07].

Note that the distance is invariant under a unitary transformation. Thus, using another coordinate system $\{\phi_j\}$ in place of $\{\psi_i\}$ would not change the distance.

Let $\{\phi_j\}$, $j = 1, 2, \ldots$ be another set of ONB functions instead of $\{\psi_i(\xi)\}$, $i = 1, 2, \ldots$. Let y_j^α be the component of f^α in $\{\phi_j\}$ where it can be expressed in terms of x_i^α by a linear relation $y_j^\alpha = \sum_{i=1}^\infty \langle \phi_j, \psi_i \rangle x_i^\alpha = \sum_{i=1}^\infty U_{i,j} x_i^\alpha$ where $U : l^2(\mathbb{Z}) \to l^2(\mathbb{Z})$, U is a unitary operator matrix $U_{i,j} = \langle \phi_j, \psi_i \rangle = \int_a^b \phi_j^*(\xi) \psi_i(\xi) d\xi$. Also, x_i^α can be written in terms of y_j^α under the following relation: $x_i^\alpha = \sum_{j=1}^\infty \langle \psi_i, \phi_j \rangle y_j^\alpha = \sum_{j=1}^\infty U_{i,j}^{-1} y_j^\alpha$ where $U_{i,j}^{-1} = \overline{U_{i,j}}$ and $\overline{U_{i,j}} = U_{j,i}^*$

$$f^\alpha(\xi) = \sum_{i=1}^\infty x_i^\alpha(\xi) \psi_i(\xi) = \sum y_i^\alpha(\xi) \phi_i(\xi).$$

So, $U(x_i) = (y_i)$ and $\sum_{i=1}^\infty x_i^\alpha \psi_i(\xi) = \sum_{j=1}^\infty y_j^\alpha \phi_j(\xi)$

$$x_i^\alpha = \langle \psi_i, f^\alpha \rangle = \int_a^b \psi_i(\xi) f^{(\alpha)}(\xi) d\xi.$$

The squared magnitude $|x_i^{(\alpha)}|^2$ of the coefficient for ψ_i in the expansion of $f^{(\alpha)}$ can be considered as a good measure of the average in the ensemble

$$Q_i = \sum_{\alpha=1}^n w^{(\alpha)} |x_i^{(\alpha)}|^2$$

can be considered a measure of the importance of $\{\psi_i\}$.

$$Q_i \geq 0, \quad \sum_i Q_i = 1.$$

See [Wat67].

Then the entropy function in terms of the Q_i's is defined as

$$S(\{\psi_i\}) = - \sum_i Q_i \log Q_i.$$

We are interested in minimizing the entropy, i.e., if $\{\Theta_j\}$ is one such optimal coordinate system, we shall have

$$S(\{\Theta_j\}) = \min_{\{\psi_j\}} S(\{\psi_i\}).$$

Let $G(\xi, \xi') = \sum_\alpha w^\alpha f^\alpha(\xi) f^{\alpha*}(\xi')$. Then G is a Hermitian matrix and $Q_i = G(i, i) = \sum_\alpha w^\alpha x_i^\alpha x_i^{\alpha*}$ where the normalization $\sum Q_i = 1$ gives us trace $G = 1$ where the trace means the diagonal sum.

Then define a special function system $\{\Theta_k(\xi)\}$ as the set of eigenfunctions of G, i.e.,

$$\int_a^b G(\xi, \xi') \Theta_k(\xi) d\xi' = \lambda_k \Theta_k(\xi). \tag{5.7}$$

So $G\Theta_k(\xi) = \lambda_k \Theta_k(\xi)$.

When the data are not functions but vectors v^αs whose components are $x_i^{(\alpha)}$ in the ψ_i coordinate system, we have

$$\sum_{i'} G(i, i') t_{i'}^k = \lambda_k t_i^k, \tag{5.8}$$

where t_i^k is the ith component of the vector Θ_k in the coordinate system $\{\psi_i\}$. So we get $\psi : \mathcal{H} \to (x_i)$ and also $\Theta : \mathcal{H} \to (t_i)$. The two ONBs result in

$$x_i^\alpha = \sum_k c_k^\alpha t_i^k \text{ for all } i, c_k^\alpha = \sum_i t_i^{k*} x_i^\alpha$$

which is the Karhunen–Loève (KL) expansion of $f^\alpha(\xi)$ or vector v^α. Then $\{\Theta_k(\xi)\}$ is the K-L coordinate system dependent on $\{w^\alpha\}$ and $\{f^\alpha(\xi)\}$. Then we arrange the corresponding functions or vectors in the order of eigenvalues $\lambda_1 \geq \lambda_2 \geq \cdots \geq \lambda_{k-1} \geq \lambda_k \geq \cdots$.

Now, $Q_i = G_{i,i} = \langle \psi_i G \psi_i \rangle = \sum_k A_{ik} \lambda_k$ where $A_{ik} = t_i^k t_i^{k*}$, which is a double stochastic matrix. Then

$$G = U \begin{pmatrix} \lambda_1 & \cdots & 0 \\ 0 & \ddots & 0 \\ 0 & \cdots & \lambda_k \end{pmatrix} U^{-1}. \tag{5.9}$$

Example 5.1. Suppose X_t is a stochastic process indexed by t in a finite interval J, and taking values in $L^2(\Omega, P)$ for some probability space (Ω, P). Assume the normalization $E(X_t) = 0$. Suppose the integral kernel $E(X_t X_s)$ can be diagonalized, i.e., suppose that

$$\int_J E(X_t X_s) \phi_k(s) ds = \lambda_k \phi_k(t)$$

with an ONB (ϕ_k) in $L^2(J)$. If $E(X_t) = 0$, then

$$X_t(\omega) = \sum_k \sqrt{\lambda_k} \phi_k(t) Z_k(\omega), \quad \omega \in \Omega,$$

where $E(Z_j Z_k) = \delta_{j,k}$, and $E(Z_k) = 0$. The ONB (ϕ_k) is called the *KL-basis* with respect to the stochastic processes $\{X_t : t \in I\}$.

The KL theorem [Ash90] states that if (X_t) is Gaussian, then so are the random variables (Z_k). Furthermore, they are $N(0,1)$, i.e., normal with mean zero and variance one, so independent and identically distributed. This last fact explains the familiar *optimality* of KL in transform coding.

Remark 5.2. Consider the case when

$$E(X_t X_s) = \frac{1}{2}(t^{2H} + s^{2H} - |t - s|^{2H})$$

and $H \in (0,1)$ is fixed. If $J = \mathbb{R}$ in the above application of KL to stochastic processes-then it is possible by a fractional integration to make the $L^2(\mathbb{R})$-ONB consist of wavelets, i.e.,

$$\psi_{j,k}(t) := 2^{j/2}\psi(2^j t - k), \quad j, k \in \mathbb{Z}, \text{ i.e., double-indexed,}$$

$$\times \, t \in \mathbb{R}, \text{ for some } \psi \in L^2(\mathbb{R}),$$

see, e.g., [Jor06a]. The process X_t is called H-fractional Brownian motion, as outlined in, e.g., [Jor06a] p. 57.

The following theorem makes clear the connection to Hilbert space geometry as used:

Theorem 5.3. *Let (Ω, P) be a probability space, $J \subset \mathbb{R}$, an interval (possibly infinite), and let $(X_t)_{t \in J}$ be a stochastic process with values in $L^2(\Omega, P)$. Assume $E(X_t) = 0$ for all $t \in J$. Then $L^2(J)$ splits as an orthogonal sum*

$$L^2(J) = \mathcal{H}_d \oplus \mathcal{H}_c \tag{5.10}$$

(d is for discrete and c is for continuous) such that the following data exists:

(a) $(\phi_k)_{k \in \mathbb{N}}$ *an ONB in $\mathcal{H}_d$.*
(b) $(Z_k)_{k \in \mathbb{N}}$: *independent random variables.*
(c) $E(Z_j Z_k) = \delta_{j,k}$, *and $E(Z_k) = 0$.*
(d) $(\lambda_k) \subset \mathbb{R}_{\geq 0}$.
(e) $\phi(\cdot, \cdot)$: *a Borel measure on $\mathbb{R}$ in the first variable, such that*

   (i) $\phi(E, \cdot) \in \mathcal{H}_c$ *for E an open subinterval of J*

   and

   (ii) $\langle \phi(E_1, \cdot) | \phi(E_2, \cdot) \rangle_{L^2(J)} = 0$ *whenever $E_1 \cap E_2 = \emptyset$.*

(f) $Z(\cdot, \cdot)$: *a measurable family of random variables such that $Z(E_1, \cdot)$ and $Z(E_2, \cdot)$ are independent when $E_1, E_2 \in \mathcal{B}_J$ and $E_1 \cap E_2 = \emptyset$,*

$$E(Z(\lambda, \cdot)Z(\lambda', \cdot)) = \delta(\lambda - \lambda'), \quad \text{and} \quad E(Z(\lambda, \cdot)) = 0.$$

Finally, we get the following Karhunen–Loève expansions for the $L^2(J)$-operator with integral kernel $E(X_t X_s)$:

$$\sum_{k\in\mathbb{N}} \lambda_k |\phi_k\rangle\langle\phi_k| + \int_J \lambda |\phi(d\lambda,\cdot)\rangle\langle\phi(d\lambda,\cdot)|. \tag{5.11}$$

Moreover, the process decomposes thus:

$$X_t(\omega) = \sum_{k\in\mathbb{N}} \sqrt{\lambda_k} Z_k(\omega)\phi_k(t) + \int_J \sqrt{\lambda} Z(\lambda,\omega)\phi(d\lambda,t). \tag{5.12}$$

Proof. By assumption, the integral operator in $L^2(J)$ with kernel $E(X_t X_s)$ is selfadjoint, positive semidefinite, but possibly unbounded. By the Spectral Theorem, this operator has the following representation:

$$\int_0^\infty \lambda Q(d\lambda),$$

where $Q(\cdot)$ is a projection-valued measure defined on the Borel subsets $\mathcal{B}$, of $\mathbb{R}_{\geq 0}$. Recall

$$Q(S_1 \cap S_2) = Q(S_1)Q(S_2) \quad \text{for } S_1, S_2 \in \mathcal{B};$$

and $\int_0^\infty Q(d\lambda)$ is the identity operator in $L^2(J)$. The two closed subspaces $\mathcal{H}_d$ and $\mathcal{H}_c$ in the decomposition (5.10) are the discrete and continuous parts of the projection value measure Q, i.e., Q is discrete (or atomic) on $\mathcal{H}_d$, and it is continuous on $\mathcal{H}_c$.

Consider first

$$Q_d(\cdot) = Q(\cdot)|_{\mathcal{H}_d}$$

and let (λ_k) be the atoms. Then for each k, the non-zero projection $Q(\{\lambda_k\})$ is a sum of rank one projections $|\phi_k\rangle\langle\phi_k|$ corresponding to a choice of ONB in the λ_k subspace. (Usually, the multiplicity is one, in which case $Q(\{\lambda_k\}) = |\phi_k\rangle\langle\phi_k|$.) This accounts for the first terms in the representations (5.11) and (5.12).

We now turn to the continuous part, i.e., the subspace $\mathcal{H}_c$, and the continuous projection-valued measure

$$Q_c(\cdot) = Q(\cdot)|_{\mathcal{H}_c}.$$

The second terms in the two formulas (5.11) and (5.12) result from an application of a disintegration theorem from [DJ07], Theorem 3.4. This theorem is applied to the measure $Q_c(\cdot)$.

We remark for clarity that the term $|\phi(d\lambda,\cdot)\rangle\langle\phi(d\lambda,\cdot)|$ under the integral sign in (5.11) is merely a measurable field of projections $P(d\lambda)$. □

Our adaptation of the spectral theorem from books in operator theory (e.g., [Jor06a]) is made with view to the application at hand, and our version of Theorem 5.3 serves to make the adaptation to how operator theory is used for time series, and for encoding. We have included it here because it isn't written precisely this way elsewhere [JS07, Son07].

5.4 Frame Bounds and Subspaces

The word "frame" in the title refers to a family of vectors in Hilbert space with basis-like properties which are made precise in Definition 5.5. We will be using entropy and information as defined classically by Shannon [Sha60], and extended to operators by von Neumann [GvN48].

The reference [Ash90] offers a good overview of the basics of both. Shannon's pioneering idea was to quantify digital "information," essentially as the negative of entropy, entropy being a measure of "disorder." This idea has found a variety of applications in both signal and image processing, and in quantum information theory, see, e.g., [Kri05]. A further recent use of entropy is in digital encoding of signals and images, compressing and quantizing digital information into a finite floating-point computer register. (Here, we use the word "quantizing" [SCE01], [Smi], [Use01] in the sense of computer science.) To compress data for storage, an encoding is used which takes into consideration probability of occurrences of the components to be quantized; and hence entropy is a gauge for the encoding [JS07, Son07].

Definition 5.4. $T \in B(\mathcal{H})$ is said to be trace class if and only if $\sum \langle \psi_i | T \psi_i \rangle$ is absolutely convergent for some ONB (ψ_i). In this case, set

$$\text{tr}(T) := \sum_i^n \langle \psi_i | T \psi_i \rangle. \tag{5.13}$$

Definition 5.5. A sequence $(h_\alpha)_{\alpha \in A}$ in $\mathcal{H}$ is called a frame if there are constants $0 < c_1 \leq c_2 < \infty$ such that

$$c_1 \|f\|^2 \leq \sum_{\alpha \in A} |\langle h_\alpha | f \rangle|^2 \leq c_2 \|f\|^2 \quad \text{for all } f \in \mathcal{H}. \tag{5.14}$$

Definition 5.6. Suppose we are given a frame operator

$$G = \sum_{\alpha \in A} w_\alpha |f_\alpha\rangle \langle f_\alpha| \tag{5.15}$$

and an ONB (ψ_i). Then for each n, the numbers

$$E_n^\psi = \sum_{\alpha \in A} w_\alpha \| f_\alpha - \sum_{i=1}^{n} \langle \psi_i | f_\alpha \rangle \psi_i \|^2 \qquad (5.16)$$

are called the error terms.

Set $L : \mathcal{H} \to l^2$,

$$L : f \mapsto (\langle h_\alpha | f \rangle)_{\alpha \in A}. \qquad (5.17)$$

Lemma 5.7. *If L is as in (5.17), then $L^* : l^2 \to \mathcal{H}$ is given by*

$$L^*((c_\alpha)) = \sum_{\alpha \in A} c_\alpha h_\alpha, \qquad (5.18)$$

where $(c_\alpha) \in l^2$; and

$$L^* L = \sum_{\alpha \in A} |h_\alpha\rangle\langle h_\alpha| \qquad (5.19)$$

Lemma 5.8. *If (f_α) are the normalized vectors resulting from a frame (h_α), i.e., $h_\alpha = \|h_\alpha f_\alpha\|$, and $w_\alpha := \|h_\alpha\|^2$, then $L^* L$ has the form (5.20).*

Proof. The desired conclusion follows from the Dirac formulas. (Please see Appendix A.) Indeed

$$|h_\alpha\rangle\langle h_\alpha| = |\|h_\alpha\| f_\alpha\rangle\langle \|h_\alpha\| f_\alpha| = \|h_\alpha\|^2 |f_\alpha\rangle\langle f_\alpha| = w_\alpha P_\alpha,$$

where P_α satisfies the two rules, $P_\alpha = P_\alpha^* = P_\alpha^2$. □

Definition 5.9. Suppose we are given $(f_\alpha)_{\alpha \in A}$, a frame, non-negative numbers $\{w_\alpha\}_{\alpha \in A}$, where A is an index set, with $\|f_\alpha\| = 1$, for all $\alpha \in A$.

$$G := \sum_{\alpha \in A} w_\alpha |f_\alpha\rangle\langle f_\alpha| \qquad (5.20)$$

is called a **frame** operator associated to (f_α).

Lemma 5.10. *Note that G is trace class if and only if $\sum_\alpha w_\alpha < \infty$; and then*

$$\mathrm{tr} G = \sum_{\alpha \in A} w_\alpha. \qquad (5.21)$$

Proof. The identity (5.21) follows from the fact that all the rank-one operators $|u\rangle\langle v|$ are trace class, with

$$\mathrm{tr}|u\rangle\langle v| = \sum_{i=1}^{n} \langle \psi_i | u \rangle \langle v | \psi_i \rangle = \langle u | v \rangle.$$

In particular, $\mathrm{tr}|u\rangle\langle u| = \|u\|^2$. □

We shall consider more general frame operators

$$G = \sum_{\alpha \in A} w_\alpha P_\alpha, \tag{5.22}$$

where (P_α) is an indexed family of projections in $\mathcal{H}$, i.e., $P_\alpha = P_\alpha^* = P_\alpha^2$, for all $\alpha \in A$. Note that P_α is trace class if and only if it is finite-dimensional, i.e., if and only if the subspace $P_\alpha \mathcal{H} = \{x \in \mathcal{H} | P_\alpha x = x\}$ is finite-dimensional.

When (ψ_i) is given, set $Q_n := \sum_{i=1}^n |\psi_i\rangle\langle\psi_i|$ and $Q_n^\perp = I - Q_n$, where I is the identity operator in $\mathcal{H}$.

Lemma 5.11.

$$E_n^\psi = \mathrm{tr}(GQ_n^\perp). \tag{5.23}$$

Proof. The proof follows from the previous facts, using that

$$\|f_\alpha - Q_n f_\alpha\|^2 = \|f_\alpha\|^2 - \|Q_n f_\alpha\|^2$$

for all $\alpha \in A$ and $n \in \mathbf{N}$. The expression (5.16) for the error term is motivated as follows. The vector components f_α in Definition 5.20 are indexed by $\alpha \in A$ and are assigned weights w_α. But rather than computing $\sum_\alpha w_\alpha$ as in Lemma 5.10, we wish to replace the vectors f_α with finite approximations $Q_n f_\alpha$, then the error term (5.16) measures how well the approximation fits the data.

Lemma 5.12. $\mathrm{tr}(GQ_n) = \sum_{\alpha \in A} w_\alpha \|Q_n f_\alpha\|^2.$

Proof.

$$\mathrm{tr}(GQ_n) = \sum_i \langle\psi_i|GQ_n\psi_i\rangle = \sum_i \langle Q_n\psi_i|GQ_n\psi_i\rangle$$

$$= \sum_{i=1}^n \langle\psi_i|G\psi_i\rangle = \sum_{\alpha \in A} w_\alpha \sum_{i=1}^n |\langle\psi_i|f_\alpha\rangle|^2 = \sum_{\alpha \in A} w_\alpha \|Q_n f_\alpha\|^2,$$

as claimed. $\qquad\square$

Proof of Lemma 5.11 *Continued.* The relative error is represented by the difference:

$$\sum_{\alpha \in A} w_\alpha - \sum_{\alpha \in A} w_\alpha \|Q_n f_\alpha\|^2 = \sum_{\alpha \in A} w_\alpha \|f_\alpha\|^2 - \sum_{\alpha \in A} w_\alpha \|Q_n f_\alpha\|^2$$

$$= \sum_{\alpha \in A} w_\alpha (\|f_\alpha\|^2 - \|Q_n f_\alpha\|^2)$$

$$= \sum_{\alpha \in A} w_\alpha \|f_\alpha - Q_n f_\alpha\|^2$$

$$= \sum_{\alpha \in A} w_\alpha \|Q_n^\perp f_\alpha\|^2 = \mathrm{tr}(GQ_n^\perp).$$ □

Definition 5.13. If G is a more general frame operator (5.22) and (ψ_i) is some ONB, we shall set $E_n^\psi := \mathrm{tr}(GQ_n^\perp)$; this is called the error sequence.

The more general case of (5.22), where

$$\mathrm{rank} P_\alpha = \mathrm{tr} P_\alpha > 1, \tag{5.24}$$

corresponds to what are called subspace frames, i.e., indexed families (P_α) of orthogonal projections such that there are $0 < c_1 \le c_2 < \infty$ and weights $w_\alpha \ge 0$ such that

$$c_1 \|f\|^2 \le \sum_{\alpha \in A} w_\alpha \|P_\alpha f\|^2 \le c_2 \|f\|^2 \tag{5.25}$$

for all $f \in \mathcal{H}$.

We now make these notions precise.

Definition 5.14. A **projection** in a Hilbert space $\mathcal{H}$ is an operator P in $\mathcal{H}$ satisfying $P^* = P = P^2$. It is understood that our projections P are orthogonal, i.e., that P is a **self-adjoint** idempotent. The orthogonality is essential because, by von Neumann, we know that there is then a 1–1 correspondence between closed subspaces in $\mathcal{H}$ and (orthogonal) projections: every closed subspace in $\mathcal{H}$ is the range of a unique projection.

We shall need the following generalization of the notion Definition 5.5 of frame.

Definition 5.15. A **fusion frame** (or subspace frame) in a Hilbert space $\mathcal{H}$ is an indexed system (P_i, w_i) where each P_i is a projection, and where (w_i) is a system of numerical weights, i.e., $w_i > 0$, such that (5.25) holds: specifically, the system (P_i, w_i) is called a fusion frame when (5.25) holds.

It is clear (see also Section 5.4) that the notion of "fusion frame" contains conventional frames Definition 5.5 as a special case.

The property (5.25) for a given system controls the weighted overlaps of the variety of subspaces $\mathcal{H}_i (:= P_i(\mathcal{H}))$ making up the system, i.e., the intersections of subspaces corresponding to different values of the index. Typically, the pairwise intersections are non-zero. The case of zero pairwise intersections happens precisely when the projections are orthogonal, i.e., when $P_i P_j = 0$ for all pairs with i and j different. In frequency analysis, this might represent orthogonal frequency bands.

When vectors in $\mathcal{H}$ represent signals, we think of bands of signals being "fused" into the individual subspaces $\mathcal{H}_i$. Further, note that for a given system of subspaces, or equivalently, projections, there may be many choices of weights consistent with (5.25): The overlaps may be controlled, or weighted, in a variety of ways. The choice of weights depends on the particular application at hand.

Theorem 5.16. *The Karhunen–Loève ONB with respect to the frame operator L^*L gives the smallest error terms in the approximation to a frame operator.*

Proof. Given the operator G which is trace class and positive semidefinite, we may apply the spectral theorem to it. What results is a discrete spectrum, with the natural order $\lambda_1 \geq \lambda_2 \geq \cdots$ and a corresponding ONB (ϕ_k) consisting of eigenvectors, i.e.,

$$G\phi_k = \lambda_k \phi_k, k \in \mathbb{N} \tag{5.26}$$

called the Karhunen–Loève data. The spectral data may be constructed recursively starting with

$$\lambda_1 = \sup_{\phi \in \mathcal{H}, \|\phi\|=1} \langle \phi | G\phi \rangle = \langle \phi_1 | G\phi_1 \rangle \tag{5.27}$$

and

$$\lambda_{k+1} = \sup_{\substack{\phi \in \mathcal{H}, \|\phi\|=1 \\ \phi \perp \phi_1, \phi_2, \ldots, \phi_k}} \langle \phi | G\phi \rangle = \langle \phi_{k+1} | G\phi_{k+1} \rangle. \tag{5.28}$$

Now an application of [AK06a]; Theorem 4.1 yields

$$\sum_{k=1}^{n} \lambda_k \geq \mathrm{tr}(Q_n^\psi G) = \sum_{k=1}^{n} \langle \psi_k | G\psi_k \rangle \quad \text{for all } n, \tag{5.29}$$

where Q_n^ψ is the sequence of projections, deriving from some ONB (ψ_i) and arranged such that

$$\langle \psi_1 | G\psi_1 \rangle \geq \langle \psi_2 | G\psi_2 \rangle \geq \cdots . \tag{5.30}$$

Hence, we are comparing ordered sequences of eigenvalues with sequences of diagonal matrix entries.

Finally, we have

$$\mathrm{tr}G = \sum_{k=1}^{\infty} \lambda_k = \sum_{k=1}^{\infty} \langle \psi_k | G\psi_k \rangle < \infty.$$

The assertion in Theorem 5.16 is the validity of

$$E_n^\phi \le E_n^\psi \qquad (5.31)$$

for all $(\psi_i) \in ONB(\mathcal{H})$, and all $n = 1, 2, ...$; and moreover, that the infimum on the RHS in (5.31) is attained for the KL-ONB (ϕ_k). But in view of our lemma for E_n^ψ, Lemma 5.11, we see that (5.31) is equivalent to the system (5.29) in the Arveson–Kadison theorem. $\square$

The Arveson–Kadison theorem is the assertion (5.29) for trace class operators, see, e.g., refs. [Arv07] and [AK06a]. That (5.31) is equivalent to (5.29) follows from the definitions.

Remark 5.17. Even when the operator G is not trace class, there is still a conclusion about the relative error estimates. With two (ϕ_i) and $(\psi_i) \in ONB(\mathcal{H})$ and with $n < m$, m large, we may introduce the following relative error terms:

$$E_{n,m}^\phi = \mathrm{tr}(G(Q_m^\phi - Q_n^\phi))$$

and

$$E_{n,m}^\psi = \mathrm{tr}(G(Q_m^\psi - Q_n^\psi)).$$

If m is fixed, we then choose a Karhunen–Loève basis (ϕ_i) for $Q_m^\psi G Q_m^\psi$ and the following error inequality holds:

$$E_{n,m}^{\phi,KL} \le E_{n,m}^\psi.$$

Our next theorem gives Karhunen–Loève optimality for sequences of entropy numbers.

Theorem 5.18. *[JS07] The Karhunen–Loève ONB gives the smallest sequence of entropy numbers in the approximation to a frame operator.*

Proof. We begin by a few facts about entropy of trace-class operators G. The entropy is defined as

$$S(G) := -\mathrm{tr}(G \log G). \qquad (5.32)$$

The formula will be used on cut-down versions of an initial operator G. In some cases, only the cut-down might be trace class. Since the Spectral

Theorem applies to G, the RHS in (5.32) is also

$$S(G) = -\sum_{k=1}^{\infty} \lambda_k \log \lambda_k. \tag{5.33}$$

For simplicity, we normalize such that $1 = \mathrm{tr}G = \sum_{k=1}^{\infty} \lambda_k$, and we introduce the partial sums

$$S_n^{KL}(G) := -\sum_{k=1}^{n} \lambda_k \log \lambda_k \tag{5.34}$$

and

$$S_n^{\psi}(G) := -\sum_{k=1}^{n} \langle \psi_k | G\psi_k \rangle \log \langle \psi_k | G\psi_k \rangle. \tag{5.35}$$

Let $(\psi_i) \in ONB(\mathcal{H})$, and set $d_k^{\psi} := \langle \psi_k | G\psi_k \rangle$; then the inequalities (5.29) take the form

$$\mathrm{tr}(Q_n^{\psi}G) = \sum_{i=1}^{n} d_i^{\psi} \leq \sum_{i=1}^{n} \lambda_i, \quad n = 1, 2, ..., \tag{5.36}$$

where as usual an ordering

$$d_1^{\psi} \geq d_2^{\psi} \geq \cdots \tag{5.37}$$

has been chosen.

Now the function $\beta(t) := t \log t$ is convex. And application of Remark 6.3 in [AK06a] then yields

$$\sum_{i=1}^{n} \beta(d_i^{\psi}) \leq \sum_{i=1}^{n} \beta(\lambda_i), \quad n = 1, 2, ... \tag{5.38}$$

Since the RHS in (5.38) is $-\mathrm{tr}(G \log G) = -S_n^{KL}(G)$, the desired inequalities

$$S_n^{KL}(G) \leq S_n^{\psi}(G), \quad n = 1, 2, ... \tag{5.39}$$

follow. That is, the KL data minimizes the sequence of entropy numbers□

5.4.1 *Supplement*

Let G, $\mathcal{H}$ be as before, $\mathcal{H}$ is an ∞-dimensional Hilbert Space $G = \sum_{\alpha} P_{\alpha}$, $\omega_{\alpha} \geq 0$, $P_{\alpha} = P_{\alpha}^* = P_{\alpha}^2$. Suppose $\dim \mathcal{H}_{\lambda_1}(G) > 0$ where $\dim \mathcal{H}_{\lambda_1}(G) =$

$\{\phi \in \mathcal{H} | G\phi = \phi\}$ and $\lambda_1 := \sup\{\langle f|Gf\rangle, f \in \mathcal{H}, \|f\| = 1\}$, then define $\lambda_2, \lambda_3, \ldots$ recursively

$$\lambda_{k+1} := \sup\{\langle f|Gf\rangle | f \perp \phi_1, \phi_2, \ldots, \phi_k\},$$

where $\dim\mathcal{H}_{\lambda_k}(G) > 0$. Set $\mathcal{K} = \bigvee_{k=1}^{\infty} \text{span}\{\phi_1, \phi_2, \ldots, \phi_k\}$. Set $\rho := \inf\{\lambda_k | k = 1, 2, \ldots\}$, then we can apply Theorems 5.16 and 5.18 to the restriction $(G - \rho I)|_{\mathcal{K}}$, i.e., the operator $\mathcal{K} \to \mathcal{K}$ given by $\mathcal{K} \ni x \longmapsto Gx - \rho x \in \mathcal{K}$.

Actually, there are two cases for G as for $G_{\mathcal{K}} := G - \rho I$: (1) compact, (2) trace class. We did (2), but we now discuss (1):

When G or $G_{\mathcal{K}}$ is given, we want to consider

$$\begin{cases} \rho = \inf\{\langle f|Gf\rangle | \|f\| = 1\} \\ \lambda_1 = \sup\{\langle f|Gf\rangle | \|f\| = 1\}. \end{cases}$$

If $G = \sum_\alpha \omega_\alpha P_\alpha$, then $\langle f|Gf\rangle = \sum_\alpha \omega_\alpha \|P_\alpha f\|^2$. If $G = \sum_\alpha |h_\alpha\rangle\langle h_\alpha|$ where $(h_\alpha) \subset \mathcal{H}$ is a family of vectors, then

$$\langle f|Gf\rangle = \sum_\alpha |\langle h_\alpha|f\rangle|^2.$$

The frame-bound condition takes the form

$$c_1\|f\|^2 \leq \sum_\alpha \omega_\alpha \|P_\alpha f\|^2 \leq c_2\|f\|^2$$

or in the standard frame case $\{h_\alpha\}_{\alpha \in A} \in FRAME(\mathcal{H})$

$$\langle f|Gf\rangle = \sum_\alpha |\langle h_\alpha|f\rangle|^2 \leq c_2\|f\|^2.$$

Lemma 5.19. *If a frame system (fusion frame or standard frame) has frame bounds $0 < c_1 \leq c_2 < \infty$, then the spectrum of the operator G is contained in the closed interval $[c_1, c_2] = \{x \in \mathbb{R} | c_1 \leq x \leq c_2\}$.*

Proof. It is clear from the formula for G that $G = G^*$. Hence, the spectrum theorem applies, and the result follows. In fact, if $z \in \mathbb{C} \setminus [c_1, c_2]$, then

$$\|zf - Gf\| \geq \text{dist}(z, [c_1, c_2])\|f\|,$$

so, $zI - G$ is invertible. So z is in the resolvent set. Hence, $\text{spec}(G) \subset [c_1, c_2]$. $\square$

A frame is a system of vectors which satisfies the two *a priori* estimates (5.14), so by the Dirac notation, it may be thought of as a statement about rank-one projections. The notion of fusion frame is the same, only with finite-rank projections; see (5.25).

5.5 Splitting Off Rank-One Operators

The general principle in frame analysis is to make a recursion which introduces rank-one operators, see Definition 5.20. The theorem we will prove accomplishes that for general class of operators in infinite-dimensional Hilbert space. Our result may also be viewed as an extension of Perron–Frobenius's theorem for positive matrices. Since we do not introduce positivity in the present section, our theorem will instead include assumptions which restrict the spectrum of the operators to which our result applies.

We now introduce a few facts about operators that will be needed in the chapter. In particular, we recall Dirac's terminology [Dir47] for rank-one operators in Hilbert space. While there are alternative notations available, Dirac's bracket terminology is especially efficient for our present considerations [JS07, Son07].

Definition 5.20. Let vectors u, $v \in \mathcal{H}$. Then

$$\langle u|v \rangle = \text{inner product} \in \mathbb{C}, \tag{5.40}$$

$$|u\rangle\langle v| = \text{rank-one operator}, \mathcal{H} \to \mathcal{H}, \tag{5.41}$$

where the operator $|u\rangle\langle v|$ acts as follows:

$$|u\rangle\langle v|w = |u\rangle\langle v|w\rangle = \langle v|w\rangle u, \quad \text{for all } w \in \mathcal{H}. \tag{5.42}$$

Theorem 5.21. *Let $\mathcal{H}$ be a generally infinite-dimensional Hilbert space, and let T be a bounded operator in $\mathcal{H}$. Under the following three assumptions, we can split off a rank-one operator from T. Specifically, assume $a \in \mathbb{C}$ satisfies:*

(1) *$0 \neq a \in spec(T)$ where $spec(T)$ denotes the spectrum of T.*
(2) *$dim R(a - T)^{\perp} = 1$ where $R(a - T)$ denotes the range of the operator $aI - T$, and $\perp$ denotes the orthogonal complement.*
(3) *$lim_{n\to\infty} a^{-n}T^n x = 0$ for all $x \in R(a - T)$.*

Then it follows that the limit exists everywhere in $\mathcal{H}$ in the strong topology of $\mathcal{B}(\mathcal{H})$. Moreover, we may pick the following representation:

$$lim_{n\to\infty} a^{-n}T^n = |\xi\rangle\langle w_1| \tag{5.43}$$

for the limiting operator on $\mathcal{H}$, where

$$\|w_1\| = 1, \quad T^*w_1 = \bar{a}w_1, \quad \langle \xi|w_1 \rangle = 1, \tag{5.44}$$

in fact

$$\xi - w_1 \in \overline{R(a - T)}, \quad \text{and} \quad T\xi = a\xi, \tag{5.45}$$

where the overbar denotes closure.

Theorem 5.21 is an analogue of the Perron–Frobenius theorem (e.g., [Jor06a]). Dictated by our applications, the present Theorem 5.21 is adapted to a different context where the present assumptions are different than those in the Perron–Frobenius theorem. We have not seen it stated in the literature in this version, and the proof (and conclusions) are different from that of the standard Perron–Frobenius theorem.

Remark 5.22. For the reader's benefit, we include the following statement of the Perron–Frobenius theorem in a formulation which makes it clear how Theorem 5.21 extends this theorem.

5.5.1 *Perron–Frobenius*

Let $d < \infty$ be given and let T be a $d \times d$ matrix with entries $T_{i,j} \geq 0$, and with positive spectral radius a. Then there is a column-vector w with $w_i \geq 0$, and a row-vector ξ such that the following conditions hold:

$$Tw = aw, \quad \xi T = a\xi, \quad \text{and} \quad \xi w = 1.$$

Proof. (of Theorem 5.21) Note that by the general operator theory, we have the following formulas:

$$R(a - T)^{\perp} = N(T^* - \bar{a}) = \{y \in \mathcal{H} | T^* y = \bar{a}y\}.$$

By assumption (2), this is a one-dimensional space, and we may pick w_1 such that $T^* w_1 = \bar{a} w_1$, and $\|w_1\| = 1$. This means that

$$\{\mathbb{C}w_1\}^{\perp} = R(a - T)^{\perp\perp} = \overline{R(a - T)}$$

is invariant under T.

As a result, there is a second bounded operator G which maps the space $\overline{R(a - T)}$ into itself, and restricts T, i.e., $T|_{\overline{R(a-T)}} = G$. Further, there is a vector $\eta^{\perp} \in (\mathbb{C}w_1)^{\perp}$ such that T has the following matrix representation:

$$
\begin{array}{cc}
\mathbb{C}w_1 & (w_1)^{\perp}
\end{array}
$$
$$
T = \left(
\begin{array}{c|c}
a & 00\cdots \\
\hline
\eta^{\perp} & G
\end{array}
\right)
\begin{array}{c}
\mathbb{C}w_1 \\
(w_1)^{\perp}
\end{array}.
$$

The entry a in the top left matrix corner represents the following operator, $sw_1 \mapsto asw_1$. The vector $\eta_{\perp}$ is fixed, and $Tw_1 = aw_1 + \eta^{\perp}$. The entry $\eta^{\perp}$ in the bottom left matrix corner represents the operator $sw_1 \mapsto s\eta^{\perp}$, or $|\eta^{\perp}\rangle\langle w_1|$.

In more detail: If Q_1 and $Q_1^{\perp} = I - Q_1$ denote the respective projections onto $\mathbb{C}w_1$ and $w_1^{\perp}$, then

$$Q_1 T Q_1 = a Q_1,$$

$$Q_1^{\perp} T Q_1 = |\eta^{\perp}\rangle\langle w_1|,$$

$$Q_1 T Q_1^{\perp} = 0, \quad \text{and}$$

$$Q_1^{\perp} T Q_1^{\perp} = G.$$

Using now assumptions in Theorem 5.21 (1) and (2), we can conclude that the operator $a - G$ is invertible with bounded inverse

$$(a - G)^{-1} : (w_1)^{\perp} \to (w_1)^{\perp}.$$

We now turn to powers of operator T. An induction yields the following matrix representation:

$$T^n = \left(\begin{array}{c|c} a^n & 00\cdots \\ \hline (a^n - G^n)(a - G)^{-1}\eta^{\perp} & G^n \end{array} \right).$$

Finally, an application of Theorem 5.21 (3) yields the following operator limit:

$$a^{-n} T^n \xrightarrow[n \to \infty]{} \left(\begin{array}{c|c} 1 & 00\cdots \\ \hline (a - G)^{-1}\eta^{\perp} & 00\cdots \end{array} \right).$$

We used that $\eta^{\perp} \in \overline{R(a - T)}$, and that

$$a^{-n}(a^n - G^n)(a - G)^{-1}\eta^{\perp} = (1 - a^{-n} G^n)(a - G)^{-1}\eta^{\perp} \xrightarrow[n \to \infty]{} (a - G)^{-1}\eta^{\perp}.$$

Further, if we set $\xi := w_1 + (a - G)^{-1}\eta^{\perp}$, then

$$T\xi = aw_1 + \eta^{\perp} + G(a - G)^{-1}\eta^{\perp} = aw_1 + a(a - G)^{-1}\eta^{\perp} = a\xi.$$

Finally, note that

$$(a - G)^{-1}\eta^{\perp} \in \overline{R(a - T)} = (w_1)^{\perp}.$$

It is now immediate from this that all of the statements in the conclusion of the theorem including Theorems (5.43), (5.44), and (5.45) are satisfied for the two vectors w_1 and ξ. $\qquad\square$

5.6 Weighted Frames and Weighted Frame Operators

In this section, we address that when frames are considered in infinite-dimensional separable Hilbert space, then the trace-class condition may not hold.

There are several remedies to this, one is the introduction of a certain weighting into the analysis. Our weighting is done as follows in the simplest case: Let $(h_n)_{n \in \mathbb{N}}$ be a sequence of vectors in some fixed Hilbert space, and suppose the frame condition from Definition 5.5 is satisfied for all $f \in \mathcal{H}$. We say that (h_n) is a frame. As in Section 5.4, we introduce the analysis operator L:

$$\mathcal{H} \ni f \longmapsto (\langle h_n | f \rangle)_n \in l^2$$

and the two operators

$$G := L^* L : \mathcal{H} \to \mathcal{H} \tag{5.46}$$

and

$$G_R := L L^* : l^2 \to l^2 \tag{5.47}$$

(the Grammian).

As noted,

$$G = \sum_{n \in \mathbb{N}} |h_n\rangle \langle h_n| \tag{5.48}$$

and G_R is matrix multiplication in l^2 by the matrix $(\langle h_i | h_j \rangle)$, i.e.,

$$l^2 \ni x = (x_i) \mapsto (G_R x) = y = (y_i),$$

where

$$y_j = \sum_i \langle h_j | h_i \rangle x_i. \tag{5.49}$$

Proposition 5.23. *Let $\{h_n\}$ be a set of vectors in a Hilbert space (infinite-dimensional, separable), and suppose these vectors form a frame with frame bounds c_1, c_2.*

(a) *Let (v_n) be a fixed sequence of scalars in l^2. Then the frame operator $G = G_v$ formed from the weighted sequence $\{v_n h_n\}$ is trace class.*

(b) *If $\sum_{n=1} |v_n|^2 = 1$, then the upper frame bound for $\{v_n h_n\}$ is also c_2.*

(c) *Pick a finite subset F of the index set, typically the natural numbers $\mathbb{N}$, and then pick (v_n) in l^2 such that $v_n = 1$ for all n in F. Then on this set F, the weighted frame agrees with the initial system of frame vectors $\{h_n\}$, and the weighted frame operator G_v is not changed on F.*

Proof. (a) Starting with the initial frame $\{h_n\}_{n\in\mathbb{N}}$, we form the weighted system $\{v_n h_n\}$. The weighted frame operator arises from applying (5.48) to this modified system, i.e.,

$$G_v = \sum_{n\in\mathbb{N}} |v_n h_n\rangle\langle v_n h_n| = \sum_{n\in\mathbb{N}} |v_n|^2 |h_n\rangle\langle h_n|. \tag{5.50}$$

Let (ϵ_n) be the canonical ONB in l^2, i.e., $(\epsilon_n)_k := \delta_{n,k}$. Then $h_n = L^*\epsilon_n$, so

$$\|h_n\| \le \|L^*\|\|\epsilon_n\| = \|L^*\| = \|L\|.$$

Now apply the trace to (5.50): Suppose $\|(v_\epsilon)\|_{l^2} = 1$. Then

$$\mathrm{tr}G_v = \sum_{n\in\mathbb{N}} |v_n|^2 \mathrm{tr}|h_n\rangle\langle h_n| = \sum_{n\in\mathbb{N}} |v_n|^2 \|h_n\|^2 \le \|L\|^2 \|(v_n)\|_{l^2}^2$$

$$= \|L\|^2 = \|LL^*\| = \|L^*L\| = \|G\| = \sup(\mathrm{spec}(G)).$$

(Note that the estimate shows more: The sum of the eigenvalues of G_v is dominated by the top eigenvalue of G.) But we recall Section 5.4.1 that (h_n) is a frame with frame-bounds c_1, c_2. It follows from Section 5.4.1 that $\mathrm{spec}(G) \subset [c_1, c_2]$. This holds also if c_1 is the largest lower bound in (5.14), and c_2, the smallest upper bound; i.e., the optimal frame bounds.

Hence, c_2 is the spectral radius of G, and also $c_2 = \|G\|$. The conclusion in (a)–(b) follows.

(c) The conclusion in (c) is a immediate consequence, but now

$$\mathrm{tr}G_v = \sum_{n\in\mathbb{N}} |v_n|^2 \|h_n\|^2 \le \|G\| \sum_{n\in\mathbb{N}} |v_n|^2 = c_2\Big(\#F + \sum_{n\in\mathbb{N}\setminus F} |v_n|^2\Big),$$

where $\#F$ is the cardinality of the set specified in (c). $\qquad\square$

Remark 5.24. Let $\{h_n\}$ and $(v_n) \in l^2$ be as in the proposition and let D_v be the diagonal operator with the sequence (v_n) down the diagonal. Then $G_v = L^*|D_v|^2 L$, and $G_{R_v} = D_v^* G_R D_v$; where

$$|D_v|^2 = D_v \overline{D_v} = \begin{pmatrix} |v_1|^2 & 0 & 0 & 0 & \cdots & 0 & 0 & 0 \\ 0 & |v_2|^2 & 0 & 0 & \cdots & 0 & 0 & 0 \\ 0 & 0 & \cdot & 0 & 0 & \cdots & 0 & 0 \\ 0 & 0 & 0 & \cdot & 0 & \cdots & 0 & 0 \\ 0 & 0 & 0 & 0 & \cdot & 0 & \cdots & 0 \end{pmatrix}$$

5.6.1 $\mathcal{B}(\mathcal{H}) = \mathcal{T}(\mathcal{H})^*$

The formula $\mathcal{B}(\mathcal{H}) = \mathcal{T}(\mathcal{H})^*$ summarizes the known fact [KR97] that $\mathcal{B}(\mathcal{H})$ is a Banach dual of the Banach space of all trace-class operators.

    The conditions (5.25) and (5.14) which introduce frames (both in vector form and fusion form) may be recast with the use of this duality.

Proposition 5.25. *An operator G arising from a vector system $(h_\alpha) \subset \mathcal{H}$, or from a projection system (w_α, P_α), yields a frame with frame bounds c_1 and c_2 if and only if*

$$c_1 \mathrm{tr}(\rho) \leq \mathrm{tr}(\rho G) \leq c_2 \mathrm{tr}(\rho) \tag{5.51}$$

for all positive trace-class operators ρ on $\mathcal{H}$.

Proof. Since both (5.25) and (5.14) may be stated in the form

$$c_1 \|f\|^2 \leq \langle f|Gf \rangle \leq c_2 \|f\|^2$$

and

$$tr(|f\rangle\langle f|) = \|f\|^2,$$

it is clear that (5.51) is sufficient.

    To see it is necessary, suppose (5.25) holds, and that ρ is a positive trace operator. By the spectral theorem, there is an ONB (f_i), and $\xi_i \geq 0$ such that

$$\rho = \sum_i \xi_i |f_i\rangle\langle f_i|.$$

We now use the estimates

$$c_1 \leq \langle f_i|Gf_i \rangle \leq c_2$$

in

$$\mathrm{tr}(\rho G) = \sum_i \xi_i \langle f_i|Gf_i \rangle.$$

Since $\mathrm{tr}(\rho) = \sum_i \xi_i$, the conclusion (5.51) follows.   □

Remark 5.26. Since quantum mechanical states (see [KR97]) take the form of density matrices, the proposition makes a connection between frame theory and quantum states. Recall, a density matrix is an operator $\rho \in \mathcal{T}(\mathcal{H})_+$ with $\mathrm{tr}(\rho) = 1$.

5.7 Localization

Starting with a frame $(h_n)_{n \in \mathbb{N}}$, non-zero vectors index set $\mathbb{N}$ for simplicity; see (5.14), we introduce the operators

$$G := \sum_{n \in \mathbb{N}} |h_n\rangle\langle h_n| \tag{5.52}$$

$$G_v := \sum_{n \in \mathbb{N}} |v_n|^2 |h_n\rangle\langle h_n| \quad \text{for } v \in l^2; \tag{5.53}$$

and the components

$$G_{h_n} := |h_n\rangle\langle h_n|. \tag{5.54}$$

We further note that the individual operators G_{h_n} in (5.54) are included in the l^2-index family G_v of (5.53). To see this, take

$$v = \epsilon_n = (0, 0, ..., 0, 1, 0, ...) \quad \text{where 1 is in } n\text{th place.} \tag{5.55}$$

It is immediate that the spectrum of G_{h_n} is the singleton $\|h_n\|^2$, and we may take $\|h_n\|^{-1} h_n$ as a normalized eigenvector. Hence, for the components G_{h_n}, there are global entropy considerations. Still in applications, it is the sequence of local approximations

$$\sum_{i=1}^{m} \langle \psi_i | h_n \rangle \psi_i = Q_m^\psi h_n \tag{5.56}$$

which is accessible. It is computed relative to some $ONB(\psi_i)$. The corresponding sequence of entropy numbers is

$$S_m^\psi(h_n) := -\sum_{i=1}^{m} |\langle \psi_i | h_n \rangle|^2 \log |\langle \psi_i | h_n \rangle|^2. \tag{5.57}$$

The next result shows that for every $v \in l^2$ with $\|v\|_{l^2} = 1$, the combined operator G_v always is entropy-improving in the following precise sense.

Proposition 5.27. *Consider the operators G_v and G_{h_n} introduced in (5.53) and (5.54). Suppose $v \in l^2$ satisfies $\|v\|_{l^2} = 1$. Then for every ONB (ψ_i) and for every m,*

$$S_m^\psi(G_v) \geq \sum_{n \in \mathbb{N}} |v_n|^2 S_m^\psi(G_{h_n}). \tag{5.58}$$

Proof. Let v, ψ, and m be as specified in the proposition. Introduce the convex function $\beta(t) := t \log t$, $t \in [0, 1]$ with the convention that $\beta(0) = \beta(1) = 0$. Then

$$-S_m^\psi(G_v) = \sum_{i=1}^{m} \beta(|\langle\psi_i|G_v\psi_i\rangle|^2) \le \sum_{i=1}^{m} \sum_{n \in \mathbb{N}} |v_n|^2 \beta(|\langle\psi_i|G_{h_n}\psi_i\rangle|^2)$$

$$= \sum_{n \in \mathbb{N}} |v_n|^2 \sum_{i=1}^{m} \beta(|\langle\psi_i|G_{h_n}\psi_i\rangle|^2) = -\sum_{n \in \mathbb{N}} |v_n|^2 S_m^\psi(G_{h_n}),$$

where we used that β is convex. In the last step, formula (5.57) was used. This proves (5.58) in the proposition. $\qquad\square$

5.8 Engineering Applications

In wavelet image compression, wavelet decomposition is performed on a digital image. Here, an image is treated as a matrix of functions where the entries are pixels. The following is an example of a representation for a digitized image function:

$$\mathbf{f(x, y)} = \begin{pmatrix} f(0,0) & f(0,1) & \cdots & f(0, N-1) \\ f(1,0) & f(1,1) & \cdots & f(1, N-1) \\ \vdots & \vdots & \vdots & \vdots \\ f(M-1,0) & f(M-1,1) & \cdots & f(M-1, N-1) \end{pmatrix}. \quad (5.59)$$

After the decomposition, quantization is performed on the image. The quantization may be a lossy (meaning some information is being lost) or lossless. Then a lossless means of compression, entropy encoding, is done on the image to minimize the memory space for storage or transmission. Here, the mechanism of entropy will be discussed.

5.8.1 *Entropy encoding*

In most images, their neighboring pixels are correlated and thus contain redundant information. Our task is to find a less correlated representation of the image, then perform redundancy reduction and irrelevancy reduction. Redundancy reduction removes duplication from the signal source (for instance, a digital image). Irrelevancy reduction omits parts of the signal that will not be noticed by the Human Visual System (HVS).

Entropy encoding further compresses the quantized values in a lossless manner which gives better compression overall. It uses a model to accurately determine the probabilities for each quantized value and produces an appropriate code based on these probabilities so that the resultant output code stream will be smaller than the input stream.

5.8.1.1 *Some terminology*

(i) Spatial Redundancy refers to correlation between neighboring pixel values.
(ii) Spectral Redundancy refers to correlation between different color planes or spectral bands.

5.8.2 *The algorithm*

Our aim is to reduce the number of bits needed to represent an image by removing redundancies as much as possible.

The algorithm for entropy encoding using Karhunen–Loève expansion can be described as follows:

1. Perform the wavelet transform for the whole image (i.e., wavelet decomposition).
2. Perform quantization to all coefficients in the image matrix, except the average detail.
3. Subtract the mean: Subtract the mean from each of the data dimensions. This produces a data set whose mean is zero.
4. Compute the covariance matrix

$$\text{cov}(X, Y) = \frac{\sum_{i=1}^{n}(X_i - \bar{X})(Y_i - \bar{Y})}{n}.$$

5. Compute the eigenvectors and eigenvalues of the covariance matrix.
6. Choose components and form a feature vector (matrix of vectors),

$$(\text{eig}_1, ..., \text{eig}_n).$$

Eigenvectors are listed in decreasing order of the magnitude of their eigenvalues. Eigenvalues found in step 5 are different in values. The eigenvector with highest eigenvalue is the principle component of the data set.

7. Derive the new data set.

$$\text{Final Data} = \text{Row Feature Matrix} \times \text{Row Data Adjust}.$$

Row Feature Matrix is the matrix that has the eigenvectors in its rows with the most significant eigenvector (i.e., with the greatest eigenvalue) at the top row of the matrix. Row Data Adjust is the matrix with mean-adjusted data transposed. That is, the matrix contains the data items in each column with each row having a separate dimension [Smi].

Please see the details in Sections 4.3.1 and 2.3.3 for the details of wavelet image decomposition, quantization, and thresholding.

Starting with a matrix representation for a particular image, we then compute the covariance matrix using the steps from (3) and (4) in the algorithm above. Next, we compute the Karhunen–Loève eigenvalues. As usual, we arrange the eigenvalues in decreasing order. The corresponding eigenvectors are arranged to match the eigenvalues with multiplicity. The eigenvalues mentioned here are the same eigenvalues in Theorems 5.16 and 5.18, thus yielding smallest error and smallest entropy in the computation.

The Karhunen–Loève transform or Principal Components Analysis (PCA) allows us to better represent each pixel on the image matrix with the smallest number of bits. It enables us to assign the smallest number of bits for the pixel that has the highest probability, then the next number to the pixel value that has second highest probability, and so forth; thus, the pixel that has the smallest probability gets assigned the highest value among all the other pixel values.

An example with letters in the text would better depict how the mechanism works. Suppose we have a text with letters a, e, f, q, r with the following probability distribution:

Letter	Probability
a	0.3
e	0.2
f	0.2
q	0.2
r	0.1

Shannon–Fano entropy encoding algorithm is outlined as follows:

- List all letters, with their probabilities in decreasing order of their probabilities.
- Divide the list into two parts with approximately equal probability (i.e., the total of probabilities of each part sums up to approximately 0.5).
- For the letters in the first part, start the code with a 0 bit and for those in the second part, with a 1.
- Recursively continue until each subdivision is left with just one letter [BCW90].

Then applying the Shannon–Fano entropy encoding scheme on the above table gives us the following assignment.

Letter	Probability	Code
a	0.3	00
e	0.2	01
f	0.2	100
q	0.2	101
r	0.1	110

Note that instead of using 8 bits to represent a letter, 2 or 3 bits are being used to represent the letters in this case.

A concrete example using a matrix and an image using principal component analysis is given in Section 5.9.

5.8.3 *Benefits of entropy encoding*

One might think that the quantization step suffices for compression. It is true that the quantization does compress the data tremendously. After the quantization step, many of the pixel values are either eliminated or replaced with other suitable values. However, those pixel values are still represented with either 8 or 16 bits. See Section 5.1.1. So we aim to minimize the number of bits used by means of entropy encoding. Karhunen–Loève transform or PCA makes it possible to represent each pixel on the digital image with the least bit representation according to their probability, thus yielding the lossless optimized representation using least amount of memory.

5.9 Principal Component Analysis

In this section, we show an illustration of how Karhunen–Loève transform or PCA is used to decompose a digital image as principal components in accordance with eigenvalues. In Section 5.3 above, we discussed the mathematical background of Karhunen–Loève transform. Please see [MP05, Mar14, KJST] for more details.

5.9.1 *The algorithm for a digital image application*

Our aim is to reduce the number of bits needed to represent an image by removing redundancies as much as possible. Karhunen–Loève transform or PCA is a transform of m vectors with the length n formed into

m-dimensional matrix $X = [X_1, \ldots, X_m]$ into a matrix Y according to

$$Y = A(X - m_X), \tag{5.60}$$

where matrix A is obtained by eigenvectors of the covariance matrix C as in (5.62) below.

The algorithm for Karhunen–Loève transform or PCA can be described as follows:

1. Take an image or data matrix X, and compute the mean of the column vectors of X, as follows:

$$m_X = E(X) = \frac{1}{n} \sum_{i=1}^{n} X_i. \tag{5.61}$$

2. Subtract the mean: Subtract the mean, m_X in (5.61) from each column vector of X. This produces a data set matrix B whose mean is zero, and it is called centering the data.

3. Compute the covariance matrix from the matrix in the previous step

$$C = \text{cov}(X) = E\left((X - m_X)(X - m_X)^T\right) \tag{5.62}$$

$$= \frac{1}{n} \sum_{i=1}^{n} X_i X_i^T - m_X m_X^T.$$

Here, $X - m_X$ can be interpreted as subtracting m_X from each column of X. $C(i, i)$ lying in the main diagonal are the variances of

$$C(i, i) = E((X_i - m_{X_i})^2). \tag{5.63}$$

Also, $C(i, j) = E\left((X_i - m_{X_i})(X_j - m_{X_j})\right)$ is the covariance between X_i and X_j.

4. Compute the eigenvectors and eigenvalues, λ_i of the covariance matrix.

5. Choose components and form a feature vector (matrix of vectors),

$$A = (\text{eig}_1, \ldots, \text{eig}_n). \tag{5.64}$$

List the eigenvectors in decreasing order of the magnitude of their eigenvalues. This matrix A is called the row feature matrix. By normalizing the column vectors of matrix A, this new matrix P becomes an orthogonal matrix. Eigenvalues found in step 4 are different in values. The eigenvector with highest eigenvalue is the principle component of the data set. Here, the eigenvectors of eigenvalues that are not up to certain specific values can be dropped, thus creating a data matrix with lower dimension value.

6. Derive the new data set.

$$\text{Final Data} = \text{Row Feature Matrix} \times \text{Row Data Adjust.}$$

The rows of the feature matrix A are orthogonal, so the inversion of PCA can be done on Equation (5.60) by

$$X = A^T Y + m_X. \qquad (5.65)$$

With the l largest eigenvalues with more variance are used instead of n eigenvalues, the matrix A_l is formed using the l corresponding eigenvectors. This yields the newly constructed data or image X' as follows:

$$X' = A_l^T Y + m_X. \qquad (5.66)$$

Row Feature Matrix is the matrix that has the eigenvectors in its rows with the most significant eigenvector (i.e., with the greatest eigenvalue) at the top row of the matrix. Row Data Adjust is the matrix with mean-adjusted data transposed. That is, the matrix contains the data items in each column with each row having a separate dimension (see, e.g., [MP05, AW12, Mar14]).

This algorithm is versatile. It can be used for a diverse set of linear data dimension reduction-problems, and it is illustrated in Figure 5.3 (see, e.g., [MP05, AW12, Mar14, KJST]).

In PCA, image compression occurs by the method of dimension reduction. Here, we need to determine how to choose the right axes. PCA gives a linear subspace of dimension that is lower than the dimension of the original image data in such a way that the image data points lie mainly in the linear subspace with the lower dimension. PCA creates a new feature space (subspace) that captures as much variance in the original image data as possible. The linear subspace is spanned by the orthogonal vectors that form a basis. These orthogonal vectors give principal axes, i.e., directions in the data with the largest variations. As in Section 5.9.1, the PCA algorithm performs the centering of the image data by subtracting off the mean, then determines the direction with the largest variation of the data and chooses an axis in that direction, and then further explores the remaining variation and locates another axis that is orthogonal to the first, and then explores as much of the remaining variation as possible. This iteration is performed until all possible axes are exhausted. Once we have a principal axis, we subtract the variance along this principal axis to obtain the remaining variance. Then the same procedure is applied again to obtain the next principal axis from the residual variance. In addition to being the direction of maximum variance, the next principal axis must be orthogonal to the other principal axes. When all the principal axes are obtained, the data set is projected onto these axes. These new orthogonal coordinate axes are also called principal components.

The outcome is all the variations along the axes of the coordinate set, and this makes the covariance matrix diagonal, which means each new variable is uncorrelated with the rest of the variables except itself. As for some of the axes that are obtained toward the last, them have very little variation. So they don't contribute much, thus, can be discarded without affecting the variability in the image data, hence reducing the dimension (see, e.g., [Mar14]).

In PCA image compression, feature selection method is used where we go through the available features of an image and select useful features such as variables or predictors, i.e., **correlation of pixel values** to the output variables.

PCA removes redundancies and describes the image data with less properties in a way that it performs a linear transformation moving the original image data to a new space spanned by the principal component. This is done by constructing a new set of properties based on combination of the old properties. The properties that present low variance are considered not useful. PCA looks for properties that have maximal variation across the data to make the principal component space. The eigenvectors found in PCA algorithm are the new set of axes of the principal component. Dimension reduction occurs when the eigenvectors with more variance are chosen but those with less variance are discarded [Mar14, KJST, MP05].

5.9.2 *Principal component analysis in a digital image*

We would like to use a color digital image PCA to illustrate dimension change in this section, so we introduce a color digital image. A color digital image is read into a matrix of pixels. We would like to use Karhunen–Loève transform or PCA applied to a digital image data to illustrate dimension reduction. Here, an image is represented as a matrix of functions where the entries are pixel values. A color image has three components. Thus, a color image matrix has three of the above image pixel matrices for red, green, and blue components and they all appear black and white when viewed "individually." We begin with the following duality principle, (i) *spatial* vs. (ii) *spectral*, and we illustrate its role in the redundancy, and for correlation of variables, in the resolution-refinement algorithm for *images*. Specifically:

(i) *Spatial Redundancy*: correlation between neighboring pixel values.

(ii) *Spectral Redundancy*: correlation between different color planes or spectral bands.

We are interested in removing these redundancies using correlations.

Starting with a matrix representation for a particular image, we then compute the covariance matrix using the steps from (3) and (4) in the algorithm above. We then compute the Karhunen–Loève eigenvalues. Next, the eigenvalues are arranged in decreasing order. The corresponding eigenvectors are arranged to match the eigenvalues with multiplicity. The eigenvalues mentioned here are the same eigenvalues λ_i in step 4 above, thus yielding smallest error and smallest entropy in the computation (see, e.g., [Son07]).

Figures 5.2–5.5 show the principal components of an image in increasing eigenvalues where the original image is a color png file.

The original file is in red, green, and blue color image, which has three R, G, B color components. So if I is the original image, it can be represented as

$$I = w_1 R + w_2 G + w_3 B = f_R + f_G + f_B, \tag{5.67}$$

where w_1, w_2, and w_3 are weights which are determined for different light intensity for color, and f_R, f_G, and f_B are the three R, G, B components of the form Equation (5.59). Each matrix appears black and white when viewed individually. Now, when PCA is performed on the Equation (5.60), I is taken as Y with the pixel values as matrix entries in Equation (5.59).

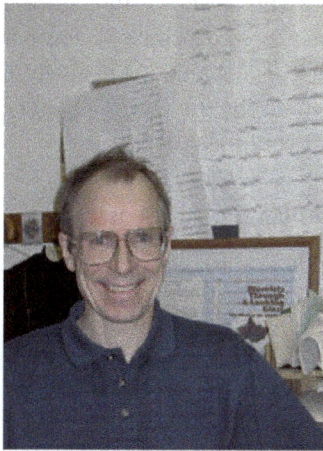

Fig. 5.2 The Original Jorgensen Image

Fig. 5.3 The three principal components of the original Jorgensen image obtained by performing PCA on each of the red, green, and blue color component of the original image, f_R, f_G and f_B. These correspond to the largest eigenvalues. From left to right: The first principal component of the red component of the image after performing PCA, the first principal component of the green component of the image after performing PCA and The first principal component of the blue component of the image after performing PCA.

Each matrix appears black and white when viewed individually. Now, it gives alternative principal components. Here, the original image I is 5.2. The original image used for 5.2, is in red, green, and blue color components which are f_R, f_G, and f_B. But PCA transform on image I will result in multiple principle components such as three principle components in 5.3.

Here, after PCA transformation, instead of RGB components three new components are used and this is shown in Figure 5.3. Note that there are more principal components. The principal components of the image are in the order of increasing eigenvalues. An image compression example is shown in Section 5.9.4 where the dimension change occurs with PCA in the direction of the principal components.

5.9.3 *A matrix example*

In this section, we provide a matrix example of PCA to illustrate how the PCA algorithm in Section 5.9.1 works.

Example 5.28. Let,

$$X = \begin{bmatrix} 1 & 0 & 0 & 3 \\ -1 & 1 & 1 & 0 \\ -1 & -2 & 4 & -5 \\ 0 & 3 & -1 & 0 \end{bmatrix} \text{ be an image matrix.} \tag{5.68}$$

Following the algorithm of Section 5.9.1, we compute the mean of each column of the above matrix, X. Then we subtract the mean of each column. For matrix X, column 1 has mean of –0.25, column 2 has mean of 0.5, column 3 has mean of 1, and column 4 has mean of –0.5.

Next, we compute the covariance matrix of X, namely $C = \text{cov}(X)$.

$$C = \text{cov}(X) = \begin{bmatrix} 0.9167 & 0.5000 & -1.3333 & 2.5000 \\ 0.5000 & 4.3333 & -4.0000 & 3.6667 \\ -1.3333 & -4.0000 & 4.6667 & -6.0000 \\ 2.5000 & 3.6667 & -6.0000 & 11.0000 \end{bmatrix}. \tag{5.69}$$

The eigenvalues of the covariance matrix $C = \text{cov}(X)$ are $\lambda_1 = 0$, $\lambda_2 = 0.3551$, $\lambda_3 = 3.2692$, and $\lambda_4 = 17.2924$. The corresponding eigenvectors of the covariance matrix, $C = \text{cov}(X)$ are as follows:

For $\lambda_1 = 0$, the corresponding eigenvector is

$$v_1 = \begin{bmatrix} 0.4295 \\ 0.5154 \\ 0.7302 \\ 0.1289 \end{bmatrix}.$$

For $\lambda_2 = 0.3551$, the corresponding eigenvector is

$$v_2 = \begin{bmatrix} 0.8584 \\ -0.1339 \\ -0.3484 \\ -0.3519 \end{bmatrix}.$$

For $\lambda_3 = 3.2692$, the corresponding eigenvector is

$$v_3 = \begin{bmatrix} 0.2240 \\ -0.7582 \\ 0.3102 \\ 0.5279 \end{bmatrix}.$$

For $\lambda_4 = 17.2924$, the corresponding eigenvector is

$$v_4 = \begin{bmatrix} 0.1685 \\ 0.3762 \\ -0.4992 \\ 0.7622 \end{bmatrix}.$$

We then form a matrix A which consists of eigenvectors in the columns of the matrix. The eigenvectors point to the direction of the principal components represented by eigenvectors with magnitude of eigenvalues.

These λ eigenvalues are the variance of the image data and the magnitude of the eigenvector direction or principal component direction. The first column of matrix A is the eigenvector v_4 corresponding to the largest eigenvalue λ_4 and the column of matrix A is the eigenvector v_3 corresponding to the largest eigenvalue λ_3 and so on in decreasing order of eigenvalues from the largest to smallest, respectively.

$$A = \begin{bmatrix} 0.1685 & 0.2240 & 0.8584 & 0.4295 \\ 0.3762 & -0.7582 & -0.1339 & 0.5154 \\ -0.4992 & 0.3102 & -0.3484 & 0.7302 \\ 0.7622 & 0.5279 & -0.3519 & 0.1289 \end{bmatrix}. \tag{5.70}$$

Then we can put the corresponding eigenvalues as diagonal entries in the D matrix as follows:

$$D = \begin{bmatrix} 17.2924 & 0 & 0 & 0 \\ 0 & 3.2692 & 0 & 0 \\ 0 & 0 & 0.3551 & 0 \\ 0 & 0 & 0 & 0 \end{bmatrix} \tag{5.71}$$

which is the diagonal matrix of eigenvalues in Section 5.3; Equation (5.9) [Son07].

Performing Principal Component Analysis on our matrix X will result in putting the eigenvectors of $C = \text{cov}(X)$ in column vector form in decreasing order of eigenvalue. After removing the least significant eigenvector v_4, we obtain A_l matrix

$$A_l = \begin{bmatrix} 0.1685 & 0.2240 & 0.8584 \\ 0.3762 & -0.7582 & -0.1339 \\ -0.4992 & 0.3102 & -0.3484 \\ 0.7622 & 0.5279 & -0.3519 \end{bmatrix}. \tag{5.72}$$

Then using Equation (5.66), we can obtain the compressed image X'.

5.9.4 *Digital image compression using principal component analysis*

In this section, we will do PCA image compression using a black and white Barbara image using different number of principal components to reconstruct the image. For more information, please see [Tha20, Reb20].

In Table 5.1, we compare the differences of compressed Barbara images, that are reconstructed using 10, 20, 30, 40, 50, 60, 70, 80, 90, and 100 principal components, respectively, and the first column labels the

Fig. 5.4 The original Barbara image from the test image pool.

Table 5.1. Table for Barbara PCA compression using different number of principal components to reconstruct the image.

No. of PCs	CR	File Size	MSE
10	10.0392	160.642 KB	9.3646
20	5.0196	172.250 KB	5.5603
30	3.3464	175.908 KB	3.3438
40	2.5098	176.492 KB	1.9192
50	2	176.436 KB	1.0752
60	1.6678	176.949 KB	0.6133
70	1.4302	178 KB	0.3315
80	1.2518	178 KB	0.1477
90	1.1130	178 KB	0.0370
100	1	178 KB	1.0363e-10

number of principal components used to reconstruct the compressed image. The second column labeled "CR" is the compression ratio which is computed by the numeric original image data/compressed image data. The third column is "File Size" of the reconstructed compressed image. The original Barbara PNG image has file size of 186KB. Please note that the file sizes are not drastically different after using 30 principal components to reconstruct the compressed image. Since the main theme of this section is the PCA image compression using different number of principal components and the compression ratios, we didn't compute PSNR and D.

Fig. 5.5 Reconstructed images of Barbara using 10, 20, 30, 40, 50, 60, 70, 80, 90, and 100 principal components, respectively, from top to bottom, left to right.

The last column is "MSE": Mean Squared Error. Please see Section 2.5. This measures the average of the squares of the errors, e.g., the average squared difference between the estimated values and the actual value. This measures the error of the compressed image in comparison to the original image. The lower the MSE is, the lesser the error is. It can be noted that when less principal components are used to reconstruct a compressed image, the file size of the compressed image is smaller, the MSE error is larger and compression ratio is large in comparison to compressed image reconstructed using more principal components. When more principal components are used to reconstruct a compressed image, the MSE is smaller, but the file size is closer to the original image's file size, and the compression ratio is closer to 1.

In comparison to wavelet image decomposition, PCA image decomposition relies on eigenvalues of the covariance matrix of an image matrix, and the corresponding principal components. The multiresolution system of wavelet decomposition where average, horizontal, vertical, and diagonal details at different levels encompass different image details, while in PCA, the dominant features picked up by principal components are used to reconstruct the images back.

Chapter 6

Matrix Factorization and Lifting

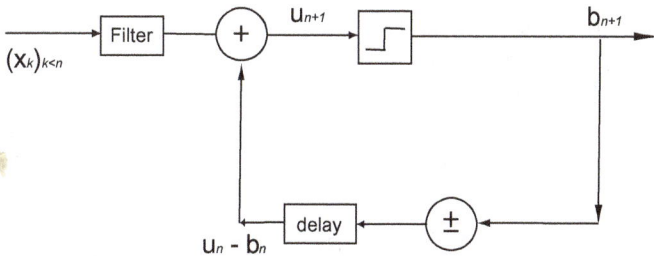

In this chapter, we discuss matrix factorizations and lifting schemes used in engineering applications that are closely related to computer engineering applications in image processing from our old paper [JS10]. Filter functions are crucial in wavelets and multiresolutions, and we explore ways to perform filter function factorization and use lifting schemes. Further results can be seen in the next chapter.

6.1 Introduction

Starting with the early work on wavelets (the 1980s), there is now an important body of theory at the crossroads of a number of mathematical areas (harmonic analysis and function theory) on the one side and theoretical signal processing on the other. An especially convincing instance of this was the recent use of wavelet algorithms by the JPEG group. (The term "JPEG" is an acronym for the Joint Photographic Experts Group which created the standard.) The achieved compression resulting from these techniques is used, e.g., in a variety of image file formats. JPEG/Exif is the most common of them used in digital cameras and other image devices. Moreover, these mathematical tools are now part of common formats for storing and transmitting photographic images on the web. The novice might find the following references helpful: [JMR01, Jor01a, Son06b, Swe96, Swe98].

The marriage of the two subjects came from the early realization that filters generating the most successful wavelet bases could be obtained from an adaptation of more classical subband filtering operations used in signal processing; with the notions of downsampling and upsampling being intrinsic to numerical wavelet algorithms used in, e.g., compression of signals and, more generally, of images. For a lucid account, see, e.g., [Law99].

A common feature for the more traditional processing tools is the division into subbands, but in modern applications such a subdivision is more subtle. Here, we develop and refine a procedure which uses factorization of families of matrix-valued functions. These operations are done on the frequency side; but it is fairly easy, at the end, to convert back to the time signal itself. Here, we use the concept of "time signal" widely allowing for systems of numbers indexed by pixels, such as gray scale numbers for still images; or, for color, more complicated configuration of pixel matrices.

To be successful, a signal processing algorithm must allow a practical procedure for breaking down an overall process into small processes. The role of factorization of matrix functions is precisely to accomplish this: In the case of two frequency bands, 2×2 matrix functions suffice and, in this case, the corresponding factorization (see, e.g., Sweldens *et al.*, [Swe96, Swe98] and [CC08]), goes by the name "lifting", and the product is a (perhaps) long string of upper and lower triangular matrices, alternating between upper and lower. But each of these basic factors will then just encode a function of the frequency variable, corresponding to a filtering step in the overall process.

In what follows, we demonstrate how this is done in the case of a process involving multiple bands, as well as the features dictated by modern applications.

6.2 Matrix-Valued Functions

Matrix-valued functions of one or more complex variables, taking values in the group SL_2, have a number of uses in both pure and applied mathematics. Here, we will focus on a framework in the signal processing literature called "the lifting scheme," or "lifting algorithms." A main result there (suitable restrictions) is the assertion that, in the case of polynomial entries, these matrix functions factor into finite products of alternating upper and lower diagonal matrix functions.

Even though pioneering ideas are from engineering, we hope to show that they are of interest in pure mathematics as well, especially in operator theory.

The result is of special practical significance in building filters using two frequency bands with a recursive input–output model; using as input filtered signals from the low band and producing an additive perturbation to the high-frequency channel. This is continued recursively, with reversal of the role of high and low in each step. For some of the literature, we refer to [SBT03, CC08, HCXH08] and many papers in the engineering literature. Since early pioneering ideas by Wim Sweldens, e.g., [SR91], the subject has since branched off in a variety of directions both applied and pure; see [DS98] and the papers cited there.

One of our motivations here is the desire to extend and refine this method to the case of multiple bands. In the simplest case, by this we mean that signals are viewed as time functions (discrete time), and each time function generates a frequency response function (generating function) of a complex variable. In applications, it is possible to encode time signals or their generating functions as vectors in a Hilbert space $\mathcal{H}$. And to do this in such a way that a finite selection of frequency bands will then correspond to a system of closed subspaces in $\mathcal{H}$. A direct generalization of the case $n = 2$ to $n > 2$ is not feasible. We note that the factorization conclusion for $n = 2$ into alternating products of upper and lower does not carry over to $n > 2$; but, motivated by applications, we outline a version that does.

A new element in our approach is the introduction of certain operator theoretic methods into the study of subband filtering in [BJ02a]; see also [BJ02b, JS09, Son06b].

While the notion of upper/lower factorization is both versatile and old, dating back to Gauss, it has a variety of modern incarnations, many of which are motivated by computation. On the pure side, we list the Iwasawa decomposition for semisimple Lie groups [Iwa49] and on the applied side, the matrix formulation of the algorithm of Gram and Schmidt for creating orthogonal vectors in complex Hilbert space [Akh65].

In signal processing, the context is different: Here, we deal with infinite-dimensional groups of matrix functions; functions taking values in one of the finite-dimensional Lie groups, different groups for different purposes.

Of the many presentations in the literature dealing with signal processing applications, the following papers are especially relevant to our present approach: [Law04, Law02, Law00, Law99, BR91, Bri10a, Bri10b, Bri13a, Bri13b]. Equally important are the papers [DS98, Swe98, Swe96]; as well as their presentation in the book [JlCH01].

6.3 Systems of Filters

In this section, we sketch the mathematics of filtering of signals (speech or images). We also endeavor to offer a dictionary translating between engineering terminology on the one hand and mathematical operations on the other.

While the method we outline applies more generally, for clarity we select our initial figures to illustrate the idea only in the simplest case. We then proceed to refinements and applications in Sections 6.5 through 6.7.

Mathematically, a discrete time signal is a numerical sequence, say (b_n) for input and (c_n) for output. The index n will typically represent time. Now the corresponding frequency response functions will be represented by a Fourier series, $g(z)$ and $h(z)$, respectively. These are functions defined on the circle group $\mathbb{T}$.

Operations on input data are represented as black boxes. Three operations enter into signal processing algorithms: (i) filter in the form of a weighted average, (ii) downsampling and (iii) upsampling. While the observed data are in the form of numerical sequences, matrix operations are more practically done on the corresponding frequency response functions.

For example, a weighted average applied to a time series turns into a Cauchy product acting on the frequency response functions. Figure 6.1 illustrates the various operators combined into a subband filtering with input and output. Specifically, in Figure 6.1, we have a fixed number N of assigned frequency subbands, so N bands in all. For each band we further have prescribed functions, so $f_0, f_1, \ldots, f_{N-1}$, with the subscript 0 indicating the lowest band. In computations, it will be handy to index the bands and the filters, by elements in the cyclic group of order $N, \{0, 1, 2, \ldots, N-1\}$, denoted $\mathbb{Z}_N$. (If $N = 2$, the bands are called low-pass and high-pass.)

In each band, as illustrated in Figure 6.1, there is a sequence of steps of operations as follows: analysis of input, filter, downsampling, upsampling, dual filter and synthesis into output. In the lowest frequency band, the filter f_0 is used and so on. In the most restricted form of signal processing, the frequency bands are arranged into orthogonal subspaces, but the orthogonality requirement will be relaxed in our present analysis. Rationale: In practice, there are better properties available!

In this section, we present the mathematics of some key concepts from signal processing. In their mathematical form, these ideas are timeless and pretty versatile; thus applying equally well to signals of a more basic nature, as well as to signal processing in wireless communication. With suitable adaptation, these in fact are tools for image processing as well.

To introduce a key idea used in what follows, consider the simplest case of a time series, here corresponding to discrete time; in other words, signals represented by sequences of complex numbers. Hence, time is represented by an arithmetic progression, here the integers $\mathbb{Z}$. As a result, the dual frequency variable will be 2π-periodic. In the language of group duality, the circle group $\mathbb{T} :=$ one torus, with Haar measure, is the dual of the additive group $\mathbb{Z}$. The introduction of group duality is helpful in several ways: it offers a compact representation of the particular Fourier analysis we need, and more importantly it offers an economical framework for analysis of frequency bands. In our following analysis, this entails representations of classical matrix groups G.

Let N be a fixed positive integer, 2 or larger. Then a subband system corresponding to N frequency bands is an N-fold partition of the period interval, or equivalently of the group T and a system of N complex-valued functions. The classical matrix groups G referred to above will now be groups of $N \times N$ complex matrices. Hence, the groups G will act on N-vectors $\mathbb{C}^N$ by matrix multiplication. We will present filter functions as functions defined on $\mathbb{T}$ and taking values in $\mathbb{C}^N$. Further, we will be using infinite-dimensional groups of functions U defined on $\mathbb{T}$ and mapping into G, referred to as matrix functions. The action of a matrix function U on a vector function F will be pointwise multiplication, i.e.,

$$(UF)(z) := U(z)F(z), \quad z \in \mathbb{T}.$$

We will further need to introduce Hilbert spaces $\mathbb{H}$ in such a manner that the groups acquire agreeable representations as they act on $\mathbb{H}$.

The purpose of our presentation here is to set up the problems for the framework of matrix analysis. By this, we mean the study of functions in one or several complex variables, but taking values in a particular Lie group of invertible matrices, e,g., the general linear group GL_N, the group $\mathcal{U}_N$ of unitary $N \times N$ matrices, or one of the symplectic groups, etc. The choice of group in our analysis depends on the problem at hand. While the Lie groups G in the above list are finite-dimensional, the moment we pass to the group of functions taking values in G, this will be an infinite-dimensional group.

Setting:

- $\mathbb{C}$: the complex plane
- $D := \{z \in \mathbb{C}; |z| < 1\}$
- $\partial D := \mathbb{T} = \{z \in \mathbb{C}; |z| = 1\} = \{e^{i\theta}; \theta \in \mathbb{R}\}$,   or $\mathbb{R}/2\pi\mathbb{Z}$
- Let $\Omega \subset \mathbb{C}$ be an open subset such that $\mathbb{T} \subset \Omega$. The Fourier representation:

$$f(z) = \sum_k a_k z^k = \sum_k a_k e^{i2\pi k\theta}. \tag{6.1}$$

If $g(z) = \sum_k b_k z^k$, we shall impose conditions at $k \to \infty$ such that

$$f(z)g(z) = \sum_n c_n z^n, \quad \text{where}$$

$$c_n = \sum_k a_k b_{n-k} \tag{6.2}$$

can be justified.

6.3.1 *Operations on time signals*

Filtering If $(b_m)_{m \in \mathbb{Z}}$ is a time signal, we say that (6.2) is a filter acting on (b_m).

In what follows, we will be using the notations $\uparrow$ and $\downarrow$ to denote operators, i.e., transformations acting on spaces of signals. These two (upsampling and downsampling) are operations on sequences [JS10].

Upsampling $\uparrow$ Fix $N \in \mathbb{Z}_+$, $N > 1$. Consider a time signal (b_k) and a frequency response function $g(z) = \sum_k b_k z^k$. Here, the function g is defined on $\mathbb{T}$, and its coefficients are the sequences used in part 1 of the definition.

Action on the signal: $(b_k) \mapsto (c_n)$, where

$$c_n = \begin{cases} b_k & \text{if } N|n, \text{i.e., } \exists k \quad \text{such that } n = N \cdot k, \\ 0 & \text{if } N \nmid n. \end{cases} \tag{6.3}$$

Action on the functions:

$$g(z) \mapsto h(z), \quad \text{where } h(z) = g(z^N). \tag{6.4}$$

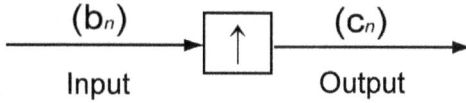

(b_n) ↑ **(c_n)**

Input   Output

Downsampling ↓ Fix $N \in \mathbb{Z}_+$, $N > 1$. Consider a time signal (b_k) and a frequency response function $g(z) = \sum_k b_k z^k$. Here, the function g is defined on $\mathbb{T}$, and its coefficients are the sequences used in part 1 of the definition. Action on the signal: $(b_k) \mapsto (c_n)$, where

$$c_n = b_{nN} \quad \text{for all } n \in \mathbb{Z} \tag{6.5}$$

(i.e., discard input b_k when k is not divisible by N). Action on the functions: $g(z) \mapsto h(z)$, where g is defined on $\mathbb{T}$ and taking values in $\mathbb{C}$ or in $\mathbb{C}^N$ and

$$h(z) = \frac{1}{N} \sum_{w \in \mathbb{T}, \; w^N = z} g(w), \quad \text{average over } N\text{th roots.} \tag{6.6}$$

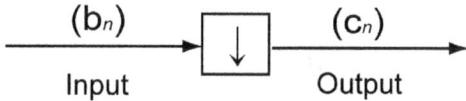

(b_n) ↓ **(c_n)**

Input   Output

Frequency bands: We say that a partition of $-\pi \le \theta < \pi$ into N sub-intervals.

$$\frac{2\pi k}{N} - \frac{\pi}{N} \le \theta < \frac{2\pi k}{N} + \frac{\pi}{N}$$

is a subband partition with $k = 0$ corresponding to the lowest band and $k = \left[\frac{N}{2}\right]$, the highest band.

Definition 6.1. Let $N \in \mathbb{Z}_+$ be given and set $\zeta_N := e^{i\frac{2\pi}{N}} = $ the principal Nth root of 1. Set

$$(A_N g)(z) := \frac{1}{N} \sum_{k=0}^{N-1} g(\zeta_N^k z). \tag{6.7}$$

Here, the two versions of the operator A_N represent transformations in sequence spaces. But by Fourier duality, this turns into associated actions on spaces of functions, so functions defined on $\mathbb{T}$.

Note the summation in (6.7) is over the cyclic group $\mathbb{Z}_N = \mathbb{Z}/N\mathbb{Z}$ $(= \{0, 1, \ldots, N-1\})$.

Lemma 6.2.

$$A_N = \uparrow\downarrow \quad \text{(downsampling followed by upsampling)}, \tag{6.8}$$

i.e., composition of operators.

Proof.

$$(A_N g)(z) = (\downarrow g)(z^N)$$
$$= \frac{1}{N} \sum_{w \in \mathbb{T}, \ w^N = z^N} g(w)$$
$$= \frac{1}{N} \sum_{k \in \mathbb{Z}_N} g(\zeta_N^k z), \quad \text{summation over the cyclic group of order } N,$$

which is the formula in (6.7). $\square$

Corollary 6.3. *The action of A_N on the time signal (b_k) is as follows:*

$$(A_N b)_k = \begin{cases} b_k & \text{if } N|k, \\ 0 & \text{if } N \nmid k. \end{cases}$$

Definition 6.4. Let $N \in \mathbb{Z}_+$ be given. If F is a vector-valued function defined on $\mathbb{T}$, by (f_k) we mean the corresponding coordinate functions. The same for G and (g_k). If f is a scalar-valued function, we denote the corresponding multiplication operator by M_f. Two systems of functions

$$F = (f_k)_{k \in \mathbb{Z}_N} \quad \text{and} \quad G = (g_k)_{k \in \mathbb{Z}_N}$$

are said to be a perfect reconstruction filter iff

$$\sum_{k \in \mathbb{Z}_N} M_{g_k} A_N M_{f_k} = I \quad \text{(see Figure 6.1)} \tag{6.9}$$

where the operator I on the RHS in (6.9) is the identity operator.

In the engineering lingo, e.g., (6.9) is expressed as follows:

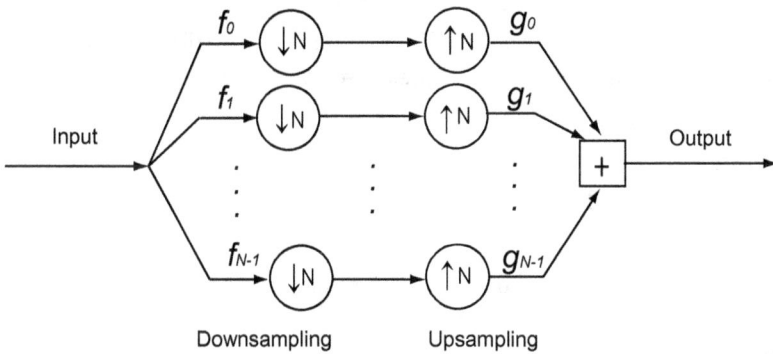

Fig. 6.1 Perfect reconstruction in subband filtering as used in signal and image processing. Input is broken down into frequency bands, processes, and then assembled (synthesis). Perfect reconstruction of output is desired.

Definition 6.5. For function $f(z) = \sum_{n \in \mathbb{Z}} a_n z^n$, set

$$f^*(z) = \sum_{n \in \mathbb{Z}} \overline{a_{-n}} z^n. \tag{6.10}$$

6.4 Groups of Matrix Functions

Groups of functions taking values in a particular Lie group G (see Section 6.2 for details) act naturally on vector-valued functions. This action is simply pointwise: If G is a group of $N \times N$ complex matrices, the action will then be on functions mapping into $\mathbb{C}^N$, i.e., complex N-space. This is important as the mathematics of filters in signal processing takes place on $\mathbb{C}^N$-valued functions. The way this is done is outlined in what follows; keeping in mind our framework of factorization for a particular (infinite-dimensional) group of functions taking values in some Lie group G [JS10].

Definition 6.6. Let F be a $\mathbb{C}^N$-valued function defined on $\mathbb{T}$ and we denote its coordinate functions (f_k) with the index k running over cyclic group $\mathbb{Z}_N$ or order N. A system $F = (f_k)_{k \in \mathbb{Z}_N}$ is said to be an orthogonal filter (with N bands) iff (6.9) holds with $g_k = f_k^*$.

Proposition 6.7. *A system* $F = (f_k)_{k \in \mathbb{Z}_N}$ *is an orthogonal filter with* N *bands iff the* $N \times N$ *matrix*

$$U_F(z) := (f_j(\zeta_N^k z))_{(j,k) \in \mathbb{Z}_N \times \mathbb{Z}_N} \tag{6.11}$$

is unitary for all $z \in \mathbb{T}(= \partial D)$.

Proof. An application of the previous lemma. □

We will consider functions on $\mathbb{T}(\subset \mathbb{C})$ taking values in (i) the scalars $\mathbb{C}$, (ii) in a fixed vector space, e.g., $\mathbb{C}^N$ for some N; or (iii) in some group of $N \times N$ complex matrices: In the first case, we write

(i) $\mathbb{T} \ni z \longmapsto f(z) \in \mathbb{C}$,
(ii) $\mathbb{T} \ni z \longmapsto F(z) = (f_k(z))_{k=1}^N \in \mathbb{C}^N$,
(iii) $\mathbb{T} \ni z \longmapsto A(z) = (A_{j,k}(z))_{j,k=1}^N$

denote a matrix function [JS10].

If A and B are matrix functions with scalar functions as entries $(A_{j,k}(z))$ and $(B_{j,k}(z))$, respectively, set $C = AB$ where $C(z) = (C_{j,k}(z))$ is the usual pointwise matrix product

$$C_{j,k}(z) = \sum_{l \in \mathbb{Z}_N} A_{j,l}(z) B_{l,k}(z).$$

Definition 6.8. An $N \times N$ matrix-valued function U is said to be unitary iff $U(z)$ is a unitary matrix for all $z \in \mathbb{T}$.

Let the set of all orthogonal N-filters be denoted $\mathcal{OF}_N$ and the set of all unitary matrix functions by $\mathcal{UM}_N$.

Definition 6.9. Let U be an $N \times N$ matrix function, and let $F = (f_k)_{k \in \mathbb{Z}_N}$ be a function system. Set

$$G(z) := U(z^N) F(z), \tag{6.12}$$

or equivalently

$$g_k(z) = \sum_{j \in \mathbb{Z}_N} U_{k,j}(z^N) f_j(z). \tag{6.13}$$

Lemma 6.10. *With the action (6.12), the group $\mathcal{UM}_F$ acts transitively on $\mathcal{OF}_N$.*

Proof. If $F \in \mathcal{OF}_N$ and $U \in \mathcal{UM}_F$, then the action (6.12) is easily seen to make $G(z) = U(z^N)F(z)$ an orthogonal filter.

Let F and G be in $\mathcal{OF}_N$ and set

$$U_{j,k}(z) = \frac{1}{N} \sum_{w \in \mathbb{T}, \ w^N = z} g_k(w) \overline{f_j(w)}. \tag{6.14}$$

An inspection shows that $U = (U_{j,k})$ is in $\mathcal{UM}_F$ and that (6.13) is satisfied. □

Corollary 6.11. *Let* $N \in \mathbb{Z}_+$ *be given. Set*

$$b(z) = \begin{bmatrix} 1 \\ z \\ z^2 \\ \vdots \\ z^{N-1} \end{bmatrix} ; \quad then \tag{6.15}$$

$$\mathcal{O}\mathcal{F}_N = \mathcal{U}\mathcal{M}_F b$$
$$=: \{U(z^N)b(z); U \in \mathcal{U}\mathcal{M}_F\}.$$

In our following analysis, we will be using more than one inner product. In a number of places, a new inner product will be defined from a given one, entailing an average over the cyclic group of order N or a system of equipartitioned points on the frequency circle. In these cases, the new inner product will be denoted $\ll \cdot, \cdot G \gg_N$ [JS10].

Definition 6.12. *Let* $N \in \mathbb{Z}_+$ *be given and let* $\langle \cdot, \cdot \rangle_N$ *be the usual inner product in* $\mathbb{C}^N$, *i.e.,*

$$\langle v, w \rangle_N := \sum_{k=1}^{N} \overline{v_k} w_k. \tag{6.16}$$

If F *and* G *are* $\mathbb{C}^N$-*valued matrix functions, set*

$$\ll F, G \gg_N (z) = \frac{1}{N} \sum_{w \in \mathbb{T}, \ w^N = z} \langle F(w), G(w) \rangle_N =\downarrow \langle F, G \rangle_N(z). \tag{6.17}$$

Lemma 6.13. *Let* $N \in \mathbb{Z}_+$ *be fixed; and let* A *and* B *be matrix functions. Then*

$$\ll Ab, Bb \gg_N = trace(A^*(z)B(z)), \tag{6.18}$$

where b is given by (6.15).

Proof.

$$\ll Ab, Bb \gg_N = \frac{1}{N} \sum_{w^N=z} \sum_j \sum_k \sum_l \overline{A_{j,k}(z)w^k} B_{j,l}(z)w^l$$

$$= \sum_j \sum_k \sum_l \overline{A_{j,k}(z)} B_{j,l}(z) \frac{1}{N} \sum_{w^N=z} \overline{w^k} w^l$$

$$= \sum_j \sum_k \sum_l \overline{A_{j,k}(z)} B_{j,l}(z) \delta_{k,l}$$

$$= \sum_j \sum_k \overline{A_{j,k}(z)} B_{j,l}(z)$$

$$= trace(A(z)^* B(z)),$$

which is the desired conclusion. $\square$

Definition 6.14. Let $\mathcal{H} = L^2(\mathbb{T})$ be the Hilbert space of functions φ given by

$$\frac{1}{2\pi} \int_{-\pi}^{\pi} |\varphi(e^{i\theta})|^2 d\theta = \sum_{n \in \mathbb{Z}} |b_n|^2, \tag{6.19}$$

where $\varphi(e^{i\theta}) = \sum_{n \in \mathbb{Z}} b_n e^{in\theta}$ is the Fourier representation of φ.

Let $F = (f_j)_{j \in \mathbb{Z}_N}$ be a function system on set

$$(S_j \varphi)(z) = f_j(z)\varphi(z^N). \tag{6.20}$$

Lemma 6.15. *Let $N \in \mathbb{Z}_+$ be given and let $F = (f_j)_{j \in \mathbb{Z}_+}$ be a function system. Then $F \in \mathcal{OF}_N$ if and only if the operators S_j in (6.20) satisfy*

$$S_j^* S_k = \delta_{j,k} I$$

$$\sum_{j \in \mathbb{Z}_N} S_j S_j^* = I,$$

where I denotes the identity operator in $\mathcal{H} = L^2(\mathbb{T})$; compare with Figure 6.1.

Proof. This is a direct application of the two previous lemmas. $\quad\square$

6.5 Group Actions

Let N be a positive integer, $N \geq 2$. A subband filter with N bands (as used in signal and image processing) is a system of functions $F := (f_0, f_1, \ldots, f_{N-1})$ defined on a frequency band, say $-\pi \leq \theta < \pi$. We will take this in the form $e^{i\theta} \in \mathbb{T}$ and then view F as a function on $\mathbb{T}$ taking values in $\mathbb{C}^N$. With Haar measure on $\mathbb{T}$, we therefore consider the Hilbert space $L^2(\mathbb{T}, \mathbb{C}^N)$ in what follows [JS10].

In this section, we outline how the entire processing system in Figure 6.1 may be encoded into a representation of a certain C^*-algebra, an algebra on N generators and two relations, called Cuntz relations, or generalized Cuntz relations. We state our first results regarding factorization in (infinite-dimensional) groups of functions taking values in some Lie group G; matrix functions for short.

We outline notational conventions and state the factorization problem in a simple case. Generalities will be added later. We begin with two key lemmas to be applied later.

Let $N \in \mathbb{Z}_+$ be given ($N > 1$) and consider $F = (f_j)_{j \in \mathbb{Z}_+}$ in $\mathcal{F}_2(N) :=$ $L^2(\mathbb{T}, \mathbb{C}^N) = \sum_0^{N-1} {}^\oplus L^2(\mathbb{T})$ where the notation in the summation symbol means orthogonal direct sum with

$$\|F\|_2^2 = \sum_{j=0}^{N-1} \|f_j\|_{L^2(\mathbb{T})}^2 < \infty.$$

We will be making use of the special vector $b \in \mathcal{F}_2(N)$,

$$b(z) = \begin{bmatrix} 1 \\ z \\ z^2 \\ \vdots \\ z^{N-1} \end{bmatrix};$$

see Corollary 6.11.

Let

$$(S_j f)(z) = z^j f(z^N) \tag{6.21}$$

be the Cuntz representation from Definition 6.14 and Lemma 6.15 [JS10].

Lemma 6.16. *Let $N \in \mathbb{Z}_+$ be fixed, $N > 1$, and let $A = (A_{j,k})$ be an $N \times N$ matrix function with $A_{j,k} \in L^2(\mathbb{T})$. Then the following two conditions are equivalent:*

(i) *For $F = (f_j) \in \mathcal{F}_2(N)$, we have $F(z) = A(z^N)b(z)$.*
(ii) *$A_{i,j} = S_j^* f_i$ where the operators S_i are from the Cuntz relations (6.21).*

Proof. (i) $\Rightarrow$ (ii). Writing out the matrix operation in (i), we get

$$f_i(z) = \sum_j A_{i,j}(z^N)z^j = \sum_j (S_j A_{i,j})(z). \tag{6.22}$$

Using $S_j^* S_k = \delta_{j,k} I$, we get $A_{i,j} = S_j^* f_i$, which is (ii).

Conversely, assuming (ii) and using $\sum_j S_i S_j^* = I$, we get $\sum_j S_j A_{i,j} = f_i$ which is equivalent to (i) by the computation in (6.22). $\square$

Corollary 6.17. *Let $N \in \mathbb{Z}_+$ be fixed, and let A and B be $N \times N$ matrix functions with L^2 entries. Then the following are equivalent:*

(i) *$A(z^N)b(z) = B(z^N)b(z)$ and*
(ii) *$A \equiv B$.*

6.5.1 *Factorizations*

We will now sketch the first step in the general conclusions about factorization.

In the following arguments, the size of the problem has two parts:

(a) The matrix size, i.e., the size of N where we consider $N \times N$ matrices.
(b) The number of factors in our factorizations.

To illustrate the idea, we begin with consideration of the case when $N = 2$ and the number of factors is also two [JS10].

Lemma 6.18. *Let*

$$A = \begin{pmatrix} \mathcal{A} & \mathcal{B} \\ \mathcal{C} & \mathcal{D} \end{pmatrix}$$

be a 2×2 matrix function and let

$$\begin{cases} f_0(z) = \mathcal{A}(z^2) + z\mathcal{B}(z^2) \\ f_1(z) = \mathcal{C}(z^2) + z\mathcal{D}(z^2). \end{cases}$$

Let L and U be scalar functions. Then the following are equivalent:

(i) $\begin{pmatrix} 1 & 0 \\ L & 1 \end{pmatrix} \begin{pmatrix} 1 & U \\ 0 & 1 \end{pmatrix} = \begin{pmatrix} \mathcal{A} & \mathcal{B} \\ \mathcal{C} & \mathcal{D} \end{pmatrix}$.

(ii) $U = S_1^* f_0$ *and* $L = S_0^* f_1$.

Proof. This is a direct consequence of the lemmas in Section 6.4. $\square$

6.5.2 *Notational conventions*

(i) Let $N \in \mathbb{Z}_+$ be fixed. We will denote $N \times N$ matrix function $A(z) = (A_{j,k}(z))_{j,k \in \mathbb{Z}_N}$ and N-column vector functions by

$$v(z) = \begin{bmatrix} v_0(z) \\ v_1(z) \\ \vdots \\ v_{N-1}(z) \end{bmatrix}.$$

We will consider A acting on the vector v as follows:

$$A_N[v](z) := A(z^N)v(z), \tag{6.23}$$

where the RHS in (6.23) is an $(N \times N)(N \times 1)$ matrix product. Note the subscript N in the definition (6.23) above.

(ii) If f and g are two scalar-valued functions, we set

$$\langle f, g \rangle_N(z) = \frac{1}{N} \sum_{\substack{w \in \mathbb{T} \\ w^N = z}} \overline{f(w)} g(w), \tag{6.24}$$

i.e., this is an inner product taking values in spaces of functions.

(iii) If f is given, we set

$$(S_f \varphi)(z) := f(z) \varphi(z^N) \tag{6.25}$$

and

$$(S_f^* \varphi)(z) := \frac{1}{N} \sum_{\substack{w \in \mathbb{T} \\ w^N = z}} \overline{f(w)} \varphi(w) = \langle f, \varphi \rangle_N(z). \tag{6.26}$$

(iv) Note that S_f^* is the $L^2(\mathbb{T})$-adjoint operator of S_j, i.e., if $\varphi, \psi \in L^2(\mathbb{T})$, then

$$\langle S_f \varphi, \psi \rangle_{L^2(\mathbb{T})} = \langle \varphi, S_f^* \psi \rangle_{L^2(\mathbb{T})}, \tag{6.27}$$

where $\langle \cdot, \cdot \rangle_{L^2(\mathbb{T})}$ denotes the usual inner product in the Hilbert space $L^2(\mathbb{T})$.

Lemma 6.19. *Let $f_0, f_1, \ldots, f_{N-1}$ be a system of N complex functions. (For the present purpose, we only need to assume that each f_j is in $L^\infty(\mathbb{T})$.) Then the following three conditions are equivalent:*

(i) *The functions f_j satisfy*

$$\langle f_j, f_k \rangle_N(z) = \delta_{j,k} 1, \quad \forall z \in \mathbb{T}, \quad \text{module orthogonality.} \tag{6.28}$$

(ii) *The operator S_{f_j} satisfies the Cuntz relations*

$$\begin{cases} S_{f_j}^* S_{f_k} = \delta_{j,k} I_{L^2(\mathbb{T})}, & \text{and} \\ \sum_{j=0}^{N-1} S_{f_j} S_{f_j}^* = I_{L^2(\mathbb{T})}. \end{cases} \tag{6.29}$$

(iii) *With $\zeta_N := e^{i\frac{2\pi}{N}}$, form the matrix function*

$$M_N(z) = (f_j(\zeta_N^k z))_{j,k \in \mathbb{Z}_N}. \tag{6.30}$$

Then M_N is a unitary matrix function.

Lemma 6.20. *The proof follows from a direct verification; see also the book [BJ02a], Chapter 2.*

Definition 6.21. A system of functions $(f_j)_{j \in \mathbb{Z}_N}$ satisfying any one of the three conditions in Lemma 6.19 is called an orthogonal system of subband filters.

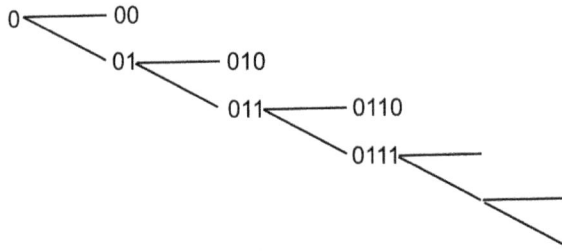

Fig. 6.2 Pyramid algorithm. Symbolic encoding of data with the use of finite words expressed in bits.

Remark 6.22. An advantage of the operator formalism in Lemma 6.19 (representations of Cuntz algebras) is that it enables the operators $P_j := S_j S_j^*$ to be a system of mutually orthogonal projections, i.e., projections onto the subspaces in $L^2(\mathbb{T}) \sim l^2(\mathbb{Z})$ corresponding to frequency bands with $P_0 =$ projection onto the subspace of the lowest band.

This representation simplifies in the case of just two bands: then the family

$$Q_i := S_1^i S_0 S_0^* S_1^{*^i}, \quad i = 0, 1, 2, \ldots$$

is infinite and mutually orthogonal. We get a well-defined infinite sum (of orthogonal projections) [JS10]:

$$\sum_{i=0}^{\infty} Q_i = I_{L^2(\mathbb{T})} \sim I_{l^2}. \tag{6.31}$$

To justify (6.31), we use that $\lim_{n \to \infty} S_1^n S_1^{*^n} = 0$ holds in the strong operator topology. With this, we then get a useful version of the pyramid algorithm, and even an image-subdivision scheme; see Figures 6.5 and 6.6.

Corollary 6.23. *Every orthogonal system of subband filters $F = [f_j]_{j \in \mathbb{Z}_N}$ has the form*

$$F = U_N[b], \tag{6.32}$$

where U is a unitary matrix function, where

$$b = \begin{pmatrix} 1 \\ z \\ z^2 \\ \vdots \\ z^{N-1} \end{pmatrix}$$

and where $U_N[b](z) = U(z^N)b(z)$.

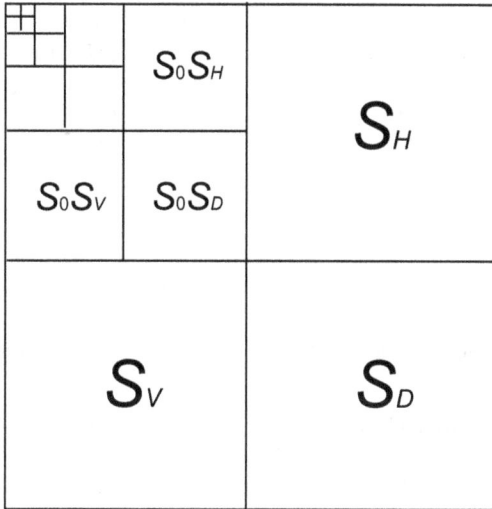

Fig. 6.3 Wavelet decomposition on an image with $N = 4$.

Definition 6.24. In the following computations, we will outline a number of finite algorithms. They will involve operations on functions. The idea is to break down operations on large systems into a sequence of steps, each of the steps acting on a smaller subsystem. The subsystems will be represented by polynomial functions, so by finite Fourier expansions.

A matrix function or a vector function is said to be of polynomial type, or a polynomial matrix function, if its entries are polynomials: If $H \subset \mathbb{Z}$ is a finite subset of the integers and $a : H \mapsto \mathbb{C}$ is a function on H, by a polynomial we shall mean the expression

$$f_H(z) := \sum_{n \in H} a_n z^n \qquad (6.33)$$

which is a finite Laurent expression.

The difference $D = \max H - \min H$ will be called the degree of f_H.

Let $N \in \mathbb{Z}_+$ be given and fixed. The following terminology will be used:

$GL_N(\text{pol})$: the $N \times N$ polynomial matrix function A such that A^{-1} is also polynomial.

$$SL_N(\text{pol}) := \{A \in GL_N(\text{pol}); \quad \det A \equiv 1\}. \qquad (6.34)$$

Our work on matrix functions gives the following:

Theorem 6.25. *(Sweldens [SR91]) Let $A \in SL_2(pol)$, then there are $l, p \in \mathbb{Z}_+$, $K \in \mathbb{C} \setminus \{0\}$ and polynomial functions $U_1, \ldots, U_p$, $L_1, \ldots, L_p$ such that*

$$A(z) = z^l \begin{pmatrix} K & 0 \\ 0 & K^{-1} \end{pmatrix} \begin{pmatrix} 1 & U_1(z) \\ 0 & 1 \end{pmatrix} \begin{pmatrix} 1 & 0 \\ L_1(z) & 1 \end{pmatrix} \cdots \begin{pmatrix} 1 & U_p(z) \\ 0 & 1 \end{pmatrix} \begin{pmatrix} 1 & 0 \\ L_p(z) & 1 \end{pmatrix}.$$

$$(6.35)$$

Remark 6.26. Note that if

$$\begin{pmatrix} \alpha & \beta \\ \gamma & \delta \end{pmatrix} \in SL_2(\text{pol}),$$

then one of the two functions $\alpha(z)$ or $\delta(z)$ must be a monomial.

6.6 Divisibility and Residues for Matrix Functions

The present section deals with some key steps in the proof of our two main theorems [JS10].

6.6.1 *The 2 × 2 case*

To highlight the general ideas, we begin with some details worked out in the 2×2 case; see Equation (7.37).

First note that from the setting in Theorem 6.25, we may assume that matrix entries have the form $f_H(z)$ as in (6.33) but with $H \subset \{0, 1, 2, \ldots\}$, i.e., $f_H(z) = a_0 + a_1 z + \cdots$. This facilitates our use of the Euclidean algorithm.

Specifically, if f and g are polynomials (i.e., $H \subset \{0, 1, 2, \ldots\}$) and if $\deg(g) \le \deg(f)$, the Euclidean algorithm yields

$$f(z) = g(z)q(z) + r(z) \tag{6.36}$$

with $\deg(r) < \deg(g)$. We shall write

$$q = \text{quot}(g, f) \quad \text{and} \quad r = \text{rem}(g, f). \tag{6.37}$$

Since

$$\begin{pmatrix} K & 0 \\ 0 & K^{-1} \end{pmatrix} \begin{pmatrix} 1 & U \\ 0 & 1 \end{pmatrix} = \begin{pmatrix} 1 & K^2 U \\ 0 & 1 \end{pmatrix} \begin{pmatrix} K & 0 \\ 0 & K^{-1} \end{pmatrix}, \tag{6.38}$$

we may assume that the factor

$$\begin{pmatrix} K & 0 \\ 0 & K^{-1} \end{pmatrix}$$

from the Equation (6.35) factorization occurs on the rightmost place.

Let U represent scalar-valued matrix entry in a matrix function. We now proceed to determine the polynomials $U_1(z), L_1(z), \ldots$, etc. inductively starting with

$$A = \begin{pmatrix} 1 & U \\ 0 & 1 \end{pmatrix} B,$$

where U and B are to be determined. Introducing (6.32), this reads

$$A(z^2) \begin{pmatrix} 1 \\ z \end{pmatrix} = \begin{pmatrix} 1 & U(z^2) \\ 0 & 1 \end{pmatrix} B(z^2) \begin{pmatrix} 1 \\ z \end{pmatrix} = \begin{pmatrix} 1 & U(z^2) \\ 0 & 1 \end{pmatrix} \begin{pmatrix} h(z) \\ k(z) \end{pmatrix}. \quad (6.39)$$

But the matrix function

$$A = \begin{pmatrix} \alpha & \beta \\ \gamma & \delta \end{pmatrix}$$

is given and fixed, see Remark 6.26. Hence,

$$\gamma(z^2) + \delta(z^2)z = k(z) \quad (6.40)$$

is also fixed. The two polynomials to be determined are u and h in (6.39). Carrying out the matrix product in (6.39) yields

$$\alpha(z^2) + \beta(z^2)z = h(z) + u(z^2)k(z) = h_0(z) + h_1(z^2)z + u(z^2)\{\gamma(z^2) + \delta(z^2)z\}$$

where we used the orthogonal splitting

$$L^2(\mathbb{T}) = S_0 S_0^* L^2(\mathbb{T}) \oplus S_1 S_1^* L^2(\mathbb{T}) \quad (6.41)$$

from Lemma 6.15. Similarly, from (6.40), we get

$$\gamma(z^2) + \delta(z^2)z = k_0(z^2) + k_1(z^2)z;$$

and therefore, $\gamma = k_0$ and $\delta = k_1$, by Lemma 6.16.

Collecting terms and using the orthogonal splitting (6.41), we arrive at the following system of polynomial equations:

$$\begin{cases} \alpha = h_0 + u\gamma \\ \beta = h_1 + u\delta; \end{cases} \quad (6.42)$$

or more precisely,

$$\begin{cases} \alpha(z) = h_0(z) + u(z)\gamma(z) \\ \beta(z) = h_1(z) + u(z)\delta(z). \end{cases}$$

It follows that the two functions u and h may be determined from the Euclidean algorithm. With (6.38), we get

$$\begin{cases} u = \text{quot}(\gamma, \alpha) \\ h_0 = \text{rem}(\gamma, \alpha) \\ h_1 = \text{rem}(\delta, \beta). \end{cases} \tag{6.43}$$

Remark 6.27. The relevance of the determinant condition we have from Theorem 6.25 is as follows:

$$\det A = \alpha\delta - \beta\gamma \equiv 1.$$

Substitution of (6.42) into this yields

$$h_0\delta - h_1\gamma \equiv 1.$$

Solutions to (6.42) are possible because the two polynomials $\delta(z)$ and $\gamma(z)$ are mutually prime. The derived matrix

$$\begin{pmatrix} h_0 & h_1 \\ \gamma & \delta \end{pmatrix}$$

is obtained from A via a row operation in the ring of polynomials.

For the inductive step, it is important to note:

$$\deg(h_0) < \deg(\gamma) \quad \text{and} \quad \deg(h_1) < \deg(\delta). \tag{6.44}$$

The next step, continuing from (6.39), is the determination of a matrix function C and three polynomials p, q, and L such that

$$\begin{pmatrix} 1 & -U \\ 0 & 1 \end{pmatrix} A = \begin{pmatrix} 1 & 0 \\ L & 1 \end{pmatrix} C \tag{6.45}$$

and

$$\begin{pmatrix} 1 & -U(z^2) \\ 0 & 1 \end{pmatrix} A(z^2) \begin{pmatrix} 1 \\ z \end{pmatrix} = \begin{pmatrix} 1 & 0 \\ L(z^2) & 1 \end{pmatrix} \begin{pmatrix} p(z) \\ q(z) \end{pmatrix}. \tag{6.46}$$

Here,

$$\begin{pmatrix} p \\ q \end{pmatrix} = C(z^2) \begin{pmatrix} 1 \\ z \end{pmatrix}.$$

The reader will note that in this step everything is as before with the only difference that now

$$\begin{pmatrix} 1 & 0 \\ L & 1 \end{pmatrix}$$

is lower diagonal in contrast with

$$\begin{pmatrix} 1 & U \\ 0 & 1 \end{pmatrix}$$

in the previous step.

This time, the determination of the polynomial p in (6.46) is automatic. With

$$p(z) = p_0(z^2) + zp_1(z^2)$$

(see (7.44)), we get the following system:

$$\begin{cases} p_0 = \alpha - u\gamma = h_0 \\ p_1 = \beta - u\delta = h_1; \quad \text{and} \end{cases}$$

$$\begin{cases} \gamma = L(\alpha - u\gamma) + q_0 = Lh_0 + q_0 \\ \delta = L(\beta - u\delta) + q_1 = Lh_1 + q_1. \end{cases}$$

So the determination of $L(z)$ and $q(z) = q_0(z^2) + zq_1(z^2)$ may be done with Euclid:

$$\begin{cases} L = \text{quot}(\alpha - u\gamma, \gamma) = \text{quot}(h_0, \gamma) \\ q_0 = \text{rem}(\alpha - u\gamma, \gamma) = \text{rem}(h_0, \gamma) \\ q_1 = \text{rem}(\beta - u\delta, \delta) = \text{rem}(h_1, \delta). \end{cases} \tag{6.47}$$

Combining the two steps, the comparison of degrees is as follows:

$$\begin{cases} \deg(q_0) < \deg(h_0) < \deg(\gamma) \\ \deg(q_1) < \deg(h_1) < \deg(\delta). \end{cases} \tag{6.48}$$

Two conclusions now follow:

(i) the procedure may continue by recursion;
(ii) the procedure must terminate.

Remark 6.28. In order to start the algorithm in (6.43) with direct reference to Euclid, we must have

$$\deg(\gamma) \le \deg(\alpha) \tag{6.49}$$

where

$$A = \begin{pmatrix} \alpha & \beta \\ \gamma & \delta \end{pmatrix}$$

is the initial 2×2 matrix function.

Now, suppose (6.49), i.e., that

$$\deg(\gamma) > \deg(\alpha).$$

Then determine a polynomial L such that

$$\deg(\gamma - L\alpha) \leq \deg(\alpha). \tag{6.50}$$

We may then start the procedure (6.43) on the matrix function

$$\begin{pmatrix} \alpha & \beta \\ \gamma - L\alpha & \delta \end{pmatrix} = \begin{pmatrix} 1 & 0 \\ -L & 1 \end{pmatrix} A.$$

If a polynomial U and a matrix function B are then found for

$$\begin{pmatrix} \alpha & \beta \\ \gamma - L\alpha & \delta \end{pmatrix},$$

then the factorization

$$A = \begin{pmatrix} 1 & 0 \\ L & 1 \end{pmatrix} \begin{pmatrix} 1 & U \\ 0 & 1 \end{pmatrix} B$$

holds; and the recursion will then work as outlined.

In the following, starting with a matrix function A, we will always assume that the degrees of the polynomials $(A_{i,j})_{i,j \in \mathbb{Z}_N}$ have been adjusted this way, so the direct Euclidean algorithm can be applied [JS10].

6.6.2 *The 3×3 case*

The thrust of this section is the assertion that Theorem 6.25 holds with small modifications in the 3×3 case.

6.6.2.1 *Comments*

In the definition of $A \in SL_3(\text{pol})$, it is understood that $A(z)$ has $\det A(z) \equiv 1$ and that the entries of the inverse matrix $A(z)^{-1}$ are again polynomials.

Note that if L, M, U, and V are polynomials, then the four matrices

$$\begin{pmatrix} 1 & 0 & 0 \\ L & 1 & 0 \\ 0 & M & 1 \end{pmatrix}, \begin{pmatrix} 1 & 0 & 0 \\ 0 & 1 & 0 \\ L & 0 & 1 \end{pmatrix}, \begin{pmatrix} 1 & U & 0 \\ 0 & 1 & V \\ 0 & 0 & 1 \end{pmatrix}, \quad \text{and} \quad \begin{pmatrix} 1 & 0 & U \\ 0 & 1 & 0 \\ 0 & 0 & 1 \end{pmatrix}$$

(6.51)

are in $SL_3(\text{pol})$ since

$$\begin{pmatrix} 1 & 0 & 0 \\ L & 1 & 0 \\ 0 & M & 1 \end{pmatrix}^{-1} = \begin{pmatrix} 1 & 0 & 0 \\ -L & 1 & 0 \\ LM & -M & 1 \end{pmatrix} \quad \text{and}$$

(6.52)

$$\begin{pmatrix} 1 & U & 0 \\ 0 & 1 & V \\ 0 & 0 & 1 \end{pmatrix}^{-1} = \begin{pmatrix} 1 & -U & UV \\ 0 & 1 & -V \\ 0 & 0 & 1 \end{pmatrix}.$$

(6.53)

Theorem 6.29. *Let $A \in SL_3(\text{pol})$; then the conclusion in Theorem 6.25 carries over with the modification that the alternating upper and lower triangular matrix functions now have the form* (6.51) *or* (6.52)–(6.53) *where the functions $L_j, M_j, U_j,$ and V_j, $j = 1, 2, \ldots$ are polynomials.*

6.6.3 *The $N \times N$ case*

Theorem 6.30. *Let $N \in \mathbb{Z}_+$, $N > 1$, be given and fixed. Let $A \in SL_N(\text{pol})$; then the conclusions in Theorem 6.25 carry over with the modification that the alternative factors in the product are upper and lower triangular matrix functions in $SL_N(\text{pol})$. We may take the lower triangular matrix factors $\mathcal{L} = (L_{i,j})_{i,j \in \mathbb{Z}_N}$ of the form*

$$\begin{pmatrix} 1 & 0 & 0 & 0 & 0 & 0 & 0 & 0 \\ 0 & 1 & 0 & 0 & 0 & 0 & 0 & 0 \\ L_p & 0 & 1 & 0 & 0 & 0 & 0 & 0 \\ 0 & L_{p+1} & 0 & 1 & 0 & 0 & 0 & 0 \\ 0 & 0 & . & 0 & 1 & 0 & 0 & 0 \\ 0 & 0 & 0 & . & 0 & 1 & 0 & 0 \\ 0 & 0 & 0 & 0 & . & 0 & 1 & 0 \\ 0 & 0 & 0 & 0 & 0 & L_{N-1} & 0 & 1 \end{pmatrix}$$

polynomial entries

$$\begin{cases} L_{i,i} \equiv 1, \\ L_{i,j}(z) = \delta_{i-j,p} L_i(z); \end{cases} \tag{6.54}$$

and the upper triangular factors of the form $\mathcal{U} = (U_{i,j})_{i,j \in \mathbb{Z}_N}$ with

$$\begin{cases} U_{i,i} \equiv 1, \\ L_{i,j}(z) = \delta_{i-j,p} U_i(z). \end{cases} \tag{6.55}$$

Proof. Notation. Let $U_1, \ldots, U_N$, $L_1, \ldots, L_N$ be polynomials and set

$$\mathcal{U}_N(U) = \begin{pmatrix} 1 & U_1 & 0 & 0 & 0 & 0 & 0 \\ 0 & 1 & U_2 & 0 & 0 & 0 & 0 \\ 0 & 0 & 1 & . & 0 & 0 & 0 \\ 0 & 0 & 0 & 1 & . & 0 & 0 \\ 0 & 0 & 0 & 0 & 1 & . & 0 \\ 0 & 0 & 0 & 0 & 0 & 1 & U_{N-1} \\ 0 & 0 & 0 & 0 & 0 & 0 & 1 \end{pmatrix} \tag{6.56}$$

$$\mathcal{L}_N(L) = \begin{pmatrix} 1 & 0 & 0 & 0 & 0 & 0 & 0 \\ L_1 & 1 & 0 & 0 & 0 & 0 & 0 \\ 0 & L_2 & 1 & 0 & 0 & 0 & 0 \\ 0 & 0 & . & 1 & 0 & 0 & 0 \\ 0 & 0 & 0 & . & 1 & 0 & 0 \\ 0 & 0 & 0 & 0 & . & 1 & 0 \\ 0 & 0 & 0 & 0 & 0 & L_{N-1} & 1 \end{pmatrix}. \tag{6.57}$$

Note that both are in $SL_N(\text{pol})$; and we have

$$\mathcal{U}_N(U)^{-1} = \mathcal{U}_N(-U) \quad \text{and}$$

$$\mathcal{L}_N(L)^{-1} = \mathcal{L}_N(-L).$$

Step 1: Starting with $A = (A_{i,j}) \in SL_N(\text{pol})$, then left-multiply with a suitably chosen $\mathcal{U}_N(-U)$ such that the degrees in the first column of $\mathcal{U}_N(-U)A$ decrease, i.e.,

$$\deg(A_{0,0}) \leq \deg(A_{1,0} - u_2 A_{1,0}) \leq \cdots \deg(A_{N-1,0}). \tag{6.58}$$

In the following, we shall use the same letter A for the modified matrix function [JS10].

Step 2: Determine a system of polynomials $L_1, \ldots, L_{N-1}$ and a polynomial vector function

$$
\begin{bmatrix} f_0 \\ f_1 \\ \ldots \\ f_{N-1} \end{bmatrix},
$$

such that,

$$
A_N \begin{bmatrix} 1 \\ z \\ z^2 \\ \ldots \\ z^{N-1} \end{bmatrix} = \mathcal{L}_N(L)_N \begin{bmatrix} f_0 \\ f_1 \\ \ldots \\ f_{N-1} \end{bmatrix}, \tag{6.59}
$$

or equivalently

$$
\sum_{j=0}^{N-1} A_{i,j}(z^N) z^j = \begin{cases} f_0(z) & \text{if } i = 0 \\ L_i(z^N) f_{i-1}(z) + f_i(z) & \text{if } i > 0 \end{cases}.
$$

Step 3: Apply the operators S_j and S_j^* from (6.21) to both sides in (6.59). First (6.59) takes the form

$$
\sum_{j=0}^{N-1} S_j A_{i,j} = \begin{cases} f_0 & \text{if } i = 0 \\ S_{f_{i-1}} L_i + f_i & \text{if } i > 0. \end{cases}
$$

For $i = 1$, we get

$$
A_{1,j} = L_1 A_{0,j} + k_j \quad \text{where } k_j = S_j^* f_1. \tag{6.60}
$$

By (6.58) and the assumptions on the matrix functions, we note that the system (6.60) may now be solved with the Euclidean algorithm

$$
\begin{cases} L_1 = \text{quot}(A_{0,j}, A_{1,j}) \\ k_j = \text{rem}(A_{0,j}, A_{1,j}) \end{cases} \tag{6.61}
$$

with the same polynomial L_1 for $j = 0, 1, \ldots, N-1$.

For the polynomial function f_1, we then have

$$
f_1 = \sum_{j=0}^{N-1} S_j k_j; \tag{6.62}
$$

i.e.,

$$f_1(z) = k_0(z^N) + k_1(z^N)z + \cdots + k_{N-1}(z^{N-1})z^{N-1}.$$

The process now continues recursively until all the functions $L_1, L_2, \ldots, f_1, f_2, \ldots$ have been determined.

Step 4: The formula (6.59) translates into a matrix factorization as follows: With L and F determined in (6.59), we get

$$A = \mathcal{L}_N(L)B \tag{6.63}$$

as a simple matrix product taking $B = (B_{i,j})$ and

$$B_{i,j} = S_j^* f_i, \tag{6.64}$$

where we used Lemmas 6.15 and 6.16.

Step 5: The process now continues with the polynomial matrix function from (6.63) and (6.64). We determine polynomials $U_1, \ldots, U_{N-1}$ and a third matrix function

$$C = (C(z)) = (C_{i,j}(z)) \quad \text{such that } B = \mathcal{U}_N(U)C.$$

Step 6: At each step of the process, we alternate L and U; and at each step, the degrees of the matrix functions are decreased. Hence, the recursion must terminate as stated in Theorem 6.30. □

6.7 Quantization

In addition to building algorithms for signal and image processing, there is the related problem of quantization. We define "quantization" broadly, and indeed there is a variety of approaches.

Indeed, the "signals" may have a subtle form; the time variable might correspond to numbers in a system of pixel grids. The tools we developed in the previous sections are sufficiently versatile. For clarity of discussion, it helps to separate quantization of the two sides, input and output; so, e.g., "time" as one and "magnitude" as the other. The idea is to select a finite set of possibilities on either side, be it points, e.g., by sampling; or one might make suitable selections of intervals on the two sides of the quantization problem.

In order to adapt to hardware and to reduce the number of computations, one selects a threshold. Specifically, when thresholding is applied to a set of numbers in an algorithm, the threshold function denoted Q in the following sends quantities under a prescribed threshold (so insignificant relative to the selected threshold) to zero.

We will continue to use the method of thresholding as above: For signals, one makes a selection of a threshold and then programs the processing to implement that quantities under the threshold are discarded.

The same idea is used in image processing. In this case, the numbers in the process will instead be grayscale pixel values. Recall that digital images are represented by a matrix of grayscale pixels. In the case of a color image, it will instead consist of three such matrices, one for each of the red, green, and blue basic components.

In the thresholding process, applied to image processing, individual pixels are marked as object pixels if their value is greater than some threshold value (assuming an object to be brighter than the background) and as background pixels otherwise; a convention known as threshold above. This contrasts with the threshold below, or threshold inside, where a pixel is labeled "object" if its value is between two thresholds; and threshold outside implies the opposite of threshold inside.

Let $x \in \mathbb{R}$ be a pixel value. An example of thresholding called hard thresholding is defined as follows:

$$T(x) = \begin{cases} 0 & \text{if } |x| \leq \lambda \\ x & \text{if } |x| > \lambda, \end{cases} \tag{6.65}$$

where $\lambda \in \mathbb{R}_+$ is the thresholding value [Wal02].

In what follows, we will outline briefly recursive quantization schemes. The purpose is to illustrate how the particular filters we developed in Section 6.5, and choice of threshold function, have the effect of making the recursive quantization schemes run faster and be more effective. A popular method in recent papers (sigma delta quantization) is based on these ideas, plus the use of subtle difference/summation algorithms, see, e.g., [LPY10, LHR09].

The literature on the subject is vast. A pioneering paper [Ben48] opens up the door to the use of spectral analysis and stochastic processes, especially amenable to the present results. On the theoretical side, recent papers are relevant: [Abd08, BOT08].

A key factor of the filtering algorithms from Sections 6.3 and 6.6 is careful use of upsampling and downsampling. With a finite filter $(h_1 h_2, \ldots)$, we get local input/output boxes (Figure 6.4):

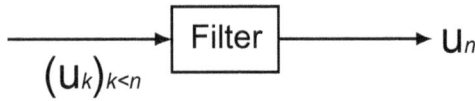

Fig. 6.4   Standard filter. Input and output represented as time series; discrete time.

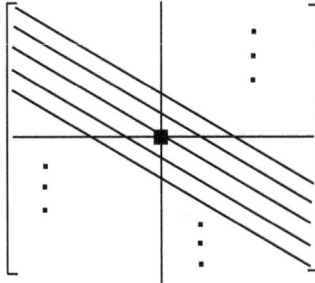

Fig. 6.5   Filter operation with slanting. See Lemma 6.19. "Decimation" entails slanted matrix representations.

$$u_n = \sum_{j\geq 1} h_j u_{n-j} = h_1 u_{n-1} + h_2 u_{n-2} + \cdots \tag{6.66}$$

or in matrix form

$$\begin{pmatrix} 0 & h_1 & h_2 & h_3 & \cdots \\ 0 & 0 & h_1 & h_2 & \cdots \\ \vdots & \vdots & \vdots & \vdots & \ddots \end{pmatrix}. \tag{6.67}$$

For contrast, compare with the standard operator matrices from (6.26)

$$\begin{pmatrix} 0 & 0 & h_1 & h_2 & h_3 & h_4 & h_5 & \cdots \\ 0 & 0 & 0 & 0 & h_1 & h_2 & h_3 & \cdots \\ 0 & 0 & 0 & 0 & 0 & 0 & h_1 & \cdots \\ \vdots & \vdots & \vdots & \vdots & \vdots & \vdots & \vdots & \cdots \\ \vdots & \vdots & \vdots & \vdots & \vdots & \vdots & \vdots & \ddots \end{pmatrix}. \tag{6.68}$$

For emphasis, we give (6.68) in diagram form (see Figure 6.5).

For the pyramid algorithm in Figure 6.2, we use two versions of the slanted matrix in Figure 6.5, high vs. low.

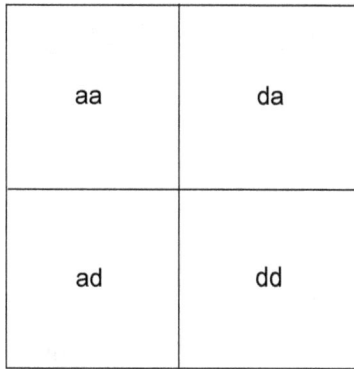

Fig. 6.6 Level 1 decomposition of an image. Clockwise: Average, horizontal, diagonal, and vertical details captured.

For the image processing (Figure 6.3), we use four versions of the slanted matrices,

(a) a matrix that takes the average in horizontal direction
(b) a matrix that takes the average in vertical direction
(c) a matrix that takes the difference in horizontal direction
(d) a matrix that takes the difference in vertical direction

which yield "average," "horizontal," "vertical", and "diagonal" details (see Figure 6.6).

We now turn to the problem of quantizing signals. These signals could be the output of a process (see Section 6.3). This step is often referred to as a case of 'Analogue to Digital' (abbreviated A/D). Quantization in our present context entails a suitable symbolic encoding, turning data from a run of a signal process (involving subband filters as in Section 6.4) into bits for subsequent computer input. Here is how Equation (6.69) illustrates this for the filters we constructed above. Quantization is essential in engineering applications; and the Q in (6.69) refers to a quantization operator.

$$\begin{cases} u_{n+1} = (Fu)_n + x_n - b_n \\ b_n = Q((Fu)_n + x_n) \end{cases} \tag{6.69}$$

Now take Figure 6.7 and Equation (6.69) together. Figure 6.7 offers a sketch of a time series as it is processed in a simple quantization filter Q, see Equation (6.69). The input is a signal x, represented as a time series and with a discrete time range k before n. So the input signal x is traced back from some fixed time n. The input is fed into one of the filters we

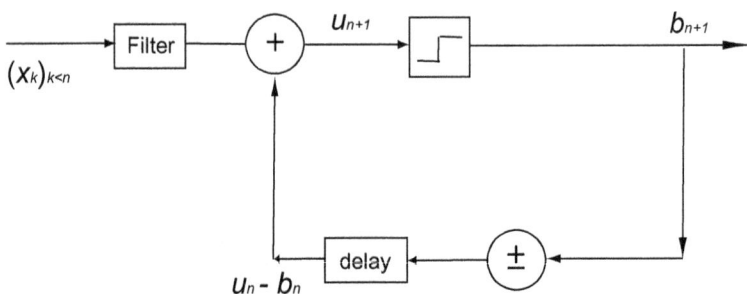

Fig. 6.7 Quantization. The operations going into a typical quantization process of a time series.

designed in Section 6.5. But now Figure 6.7 further illustrates the result of a loop for time n. The step from n to $n+1$ is spelled out in the first part in Equation (6.69) and the quantization operator Q is made precise in the second equation in (6.69). It is the loop in Figure 6.7 which involves thresholding and delay. Moving through the diagram in Figure 6.7, the next step is the process of adding the filtered signal to the output of a loop from time n. The resulting sum then contributes to time $n+1$. And the combined process is thus summarized in the discussion and in the visual in Figure 6.7.

The filter F from the first equation in (6.69) and the first box in Figure 6.7, may be any one of those built in Sections 6.4 through D.3. So the particular filter F selected may itself be the result of a factorization algorithm as outlined above: It may be a time series, a wireless signal, or a system of pixel values; and in each case, it may involve any number of frequency bands.

The output from F (see Figure 6.7) will pass through a thresholding filter Q, thus outputting b_{n+1}. In symbols, the next two steps are: "Take difference" and time-shift the result ("delay"), so from $n+1$ back to n. The first equation in (6.69) indicates how the process repeats itself, but with the output from the previous step as input in the next [JS10].

Chapter 7

Filters and Matrix Factorization

Motivated by the previous chapter and [JS10], we further discuss filters and matrix factorizations that are closely related to wavelet filters and multiresolutions, which are results from our old paper [JS14b].

7.1 Introduction

Two facts are recognized about multiresolutions — one that they are versatile, and second that they offer powerful and fast algorithms. Their use in wavelet analysis, e.g., takes advantage of a familiar localization property that wavelets have (by contrast to analogous Fourier tools). One is thus led to a host of recursive basis constructions created with the use of multiresolutions in Hilbert spaces. This philosophy carries over even to multiresolution analysis of fractals and iterated function systems in dynamics, as they too share these multiband features. Moreover, a careful use of multiresolutions and an implied localization further yields much better pointwise approximations than is otherwise possible for traditional Fourier bases. The novice might find the following references helpful: [JMR01, Jor01a].

Here, we take up a specific and algorithmic feature of this analysis: How does one effectively create an algorithm, breaking down the process into a finite number of steps, each individual step using input–output between pairs of bands in a given finite set of N frequency bands. We show that this can be accomplished with a matrix factorization; each matrix factor modeling the $(i-j)$ input–output step. While a fixed polynomial signal-processing filter, corresponding to a fixed number of N bands, will take the form of an N by N matrix having polynomial entries, the breaking down into easy implementation steps will require a relatively subtle matrix factorization; built in such a way that the matrix factors will be alternately upper- or lower triangular, and such that the factors further will then represent local $(i-j)$ input–output operations between pairs of bands in the array of a total of N bands.

While our aim is a matrix factorization covering an arbitrary number N of bands, to simplify the analysis, we begin with a detailed study of the case $N = 2$, and thus the 2 by 2 filter-matrix case. The key idea is more transparent there. We then proceed to the general case of $N > 2$, and isolating those features which require a more subtle analysis for the case $N > 2$, as compared with the $N = 2$ case.

The second step in our realization of filters requires an associated representation in a suitable Hilbert space $\mathcal{H}$ of states; designed in such a way that specific families of closed subspaces in $\mathcal{H}$ will model "non-overlapping frequency bands." Orthogonality of these subspaces (in $\mathcal{H}$) then corresponds to the desirable feature of uncorrelated frequency bands. Since the different frequency bands must exhaust the total signal for the entire system, one looks for a realization with orthogonal projections which add to the identity operator in $\mathcal{H}$. Hence, this version of time/frequency analysis is non-commutative, and one is further faced with a selection of special families of commuting orthogonal projections. We show that these may be computed from the above-mentioned matrix factorization. In an operator-algebraic language, the Hilbert space framework takes the form of suitable representations of a certain non-commutative algebra of operators; often referred to as the Cuntz algebra, and denoted $\mathcal{O}_N$. The subscript N corresponds to "number of frequency bands." The algebra $\mathcal{O}_N$ itself is specified by its generators and relations; see Sections 7.2 and 7.3. In general, $\mathcal{O}_N$ has a multitude of representations in Hilbert space. Here, we identify those of relevance to multiband signal processing.

Our purpose is to establish factorization of matrices $M_N(\mathcal{A})$ over certain rings $\mathcal{A}$ of functions, among them the ring of polynomials, and the L^∞ functions on the circle group $\mathbb{T}$. An equivalent formulation is the study of functions on $\mathbb{T}$ which take values in the $N \times N$ scalar matrices. The general setting is as follows: Fix N, and consider the group $SL_N(\mathcal{A})$ where the "S" is for determinant $= 1$. The object is then to factor arbitrary elements in $SL_N(\mathcal{A})$ as alternating products of upper and lower triangular matrix functions; equivalently, upper and lower triangular elements in $M_N(\mathcal{A})$ with the constant 1 in the diagonal.

In digital signal or image processing, one makes use of subdivisions of various families of signals in frequency bands. This is of relevance in modern-day wireless signal and image processing, and the choice of a number N of frequency bands may vary from one application to the next.

There is a certain representation theoretic framework which has proved successful: one builds a representation of the basic operations on signals, filtering, downsampling (in the complex frequency variable), upsampling, and using dual filter. These operations get represented by a system of operators in Hilbert spaces of states, say $\mathcal{H}$.

A multiresolution (see Figures 7.1 to 7.6) then takes the form of a family of closed subspaces in $\mathcal{H}$. In this construction, "non-overlapping frequency bands" correspond to orthogonal subspaces in $\mathcal{H}$; or equivalently to systems of orthogonal projections. Since the different frequency bands must exhaust the signals for the entire system, one looks for orthogonal projections which add to the identity operator in $\mathcal{H}$. This leads to the study of certain representations of the Cuntz algebra $\mathcal{O}_N$, details are given in what follows. Since time/frequency analysis is non-commutative, one is further faced with a selection of special families of commuting orthogonal projections. When these iteration schemes (repeated subdivision sequences) are applied to the initial generators, one arrives at new bases and frames; and, in other applications, to wavelet families as recursive schemes.

Our study of iterated matrix factorizations are motivated by such questions from signal processing, and arising in multiresolution analyses. In this case, elements in the group $SL_N(\mathcal{A})$ of matrix functions act on vector functions f in a complex frequency variable, where the components in f correspond to a specified system of N frequency bands. When a matrix factorization is established, then the action of the respective upper and lower triangular elements in $M_N(\mathcal{A})$ are especially simple, in that a lower triangular filter filters a low band, and then adds it to one of the higher bands; and similarly for the action of upper triangular matrix functions.

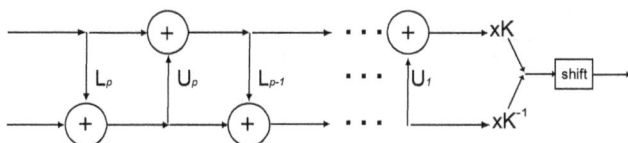

Fig. 7.1 A sketch of filter design via iterations of lower/upper matrix factorization.

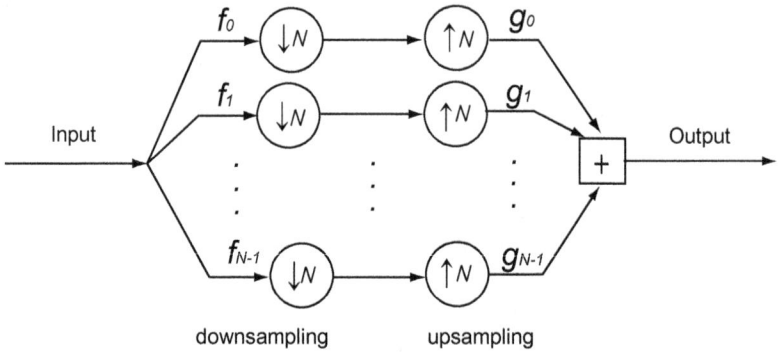

Fig. 7.2 Perfect reconstruction in subband filtering as used in signal and image processing. Input is broken down into frequency bands, processes, and then assembled (synthesis). Perfect reconstruction of output is desired.

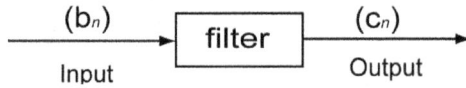

Fig. 7.3 Filtering.

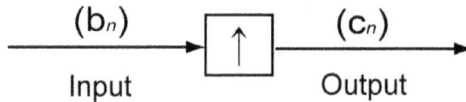

Fig. 7.4 Upsampling.

Our analysis depends on a certain representation of the Cuntz algebra $\mathcal{O}_N$, where $\mathcal{O}_N$ is an algebra generated by the basic operations on signal representations, filtering, downsampling (in the complex frequency variable), upsampling, and use of dual filter; see Figure 7.1 [JS14b].

$$\sum_n b_n z^{nN} = b_0 + b_1 z^N + b_2 z^{2N} + \cdots \ ;$$

so

$$c_n = \begin{cases} b_{n/N} & \text{if } N|n \\ 0 & \text{if } N \nmid n \end{cases}$$

$$\frac{1}{N} \sum_{w \in \mathbb{T}, w^N = z} b_n w^n = b_0 + b_N z + b_{2N} z^2 + \cdots$$

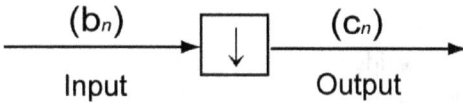

Fig. 7.5 Downsampling.

Fig. 7.6 Downsampling and upsampling.

7.2 Factorization Algorithm

In order to illustrate our use of representations of the Cuntz algebra O_N in algorithms for factorization, we begin with the case of $N = 2$. The skeleton of these algorithms has three basic steps, which we now outline.

The algorithm

Given

$$\begin{pmatrix} A & B \\ C & D \end{pmatrix} \in SL_2(\mathcal{F}),$$

where $\mathcal{F}$ is some fixed ring of functions defined on a subset $\Omega \subset \mathbb{C}$ such that $\mathbb{T} \subset \Omega$.

Step 1:

Given

$$\mathcal{A} = \begin{pmatrix} A & B \\ C & D \end{pmatrix}, \quad AD - BC \equiv 1 \quad \text{on } \mathbb{T},$$

and set

$$\mathcal{A}(z^2) \begin{pmatrix} 1 \\ z \end{pmatrix} = \begin{pmatrix} A(z^2) + zB(z^2) \\ C(z^2) + zD(z^2) \end{pmatrix}. \tag{7.1}$$

Let S_i, $i = 0, 1$ be

$$\begin{cases} S_0 f(z) = f(z^2) \\ S_1 f(z) = z f(z^2). \end{cases} \tag{7.2}$$

For the corresponding adjoint operators, we therefore get

$$\begin{cases} S_0^* f(z) = \frac{1}{2} \sum_{\omega^2 = z} f(\omega) \\ S_1^* f(z) = \frac{1}{2} \sum_{\omega^2 = z} \overline{\omega} f(\omega), \end{cases} \tag{7.3}$$

where the summation in (7.2), (7.3) are over points $z, \omega \in \mathbb{T}$.

Then $(S_i)_{i=0,1}$ are isometries in $L^2(\mathbb{T})$, and $S_i^* S_j = \delta_{i,j} I$, $\sum_{i=0}^{1} S_i S_j^* = I$ where I denotes the identity operator in the Hilbert space $L^2(\mathbb{T})$. We will want $\mathcal{F}$ to be a ring of meromorphic functions, such that they are determined by their values on $\mathbb{T} = \{z \in \mathbb{C}, |z| = 1\}$; or we are simply working with functions on $\mathbb{T}$ [JS14b].

Step 2:

Find functions L such that

$$\begin{pmatrix} 1 & 0 \\ L & 1 \end{pmatrix} \mathcal{A}_{\text{new}} = \mathcal{A}. \tag{7.4}$$

Solution: Apply (7.4) to

$$\begin{pmatrix} 1 \\ z \end{pmatrix},$$

and set

$$\mathcal{A}_{\text{new}}(z^2) \begin{pmatrix} 1 \\ z \end{pmatrix} = \begin{pmatrix} f_0(z) \\ f_1(z) \end{pmatrix};$$

then

$$\begin{cases} f_0 = A(z^2) + zB(z^2) \\ L(z^2) f_0(z) + f_1(z) = C(z^2) + zD(z^2). \end{cases} \tag{7.5}$$

Apply S_i^*, $i = 0, 1$, to (7.5)

$$\begin{cases} S_0^* f_0 = A, \quad S_1^* f_0 = B \\ L S_0^* f_0 + S_0^* f_1 = C \\ L S_1^* f_0 + S_1^* f_1 = D. \\ \Rightarrow L = \frac{C - S_0^* f_1}{A}; \quad L = \frac{D - S_1^* f_1}{B}. \end{cases} \tag{7.6}$$

Corollary 7.1. $A(S_1^* f_1) - B(S_0^* f_1) = 1.$

Proof. Consider (7.6) with $\det \mathcal{A} = 1$.

$$\mathcal{A} = \begin{pmatrix} A & B \\ C & D \end{pmatrix}$$

with $AD - BC = 1$. So

$$\begin{pmatrix} A & B \\ S_0^* f_1 & S_1^* f_1 \end{pmatrix} \in SL_2(\mathcal{F}).$$

We now assume $S_i \mathcal{F} \subset \mathcal{F}$, and $S_i^* \mathcal{F} \subset \mathcal{F}$, for all $i = 0, 1$. □

Step 3:

Having the form L, from (7.4) we get

$$\mathcal{A}_{\text{new}} = \begin{pmatrix} l & 0 \\ -L & 1 \end{pmatrix} \begin{pmatrix} A & B \\ C & D \end{pmatrix} = \begin{pmatrix} A & B \\ -LA+C & -LB+D \end{pmatrix} = \begin{pmatrix} A & B \\ S_0^* f_1 & S_1^* f_1 \end{pmatrix}.$$

Step 4:

$$\begin{pmatrix} l & U \\ 0 & 1 \end{pmatrix} \mathcal{A}_{\text{up}} = \mathcal{A}_{\text{new}}.$$

Set

$$\mathcal{A}_{\text{up}}(z^2) \begin{pmatrix} 1 \\ z \end{pmatrix} = \begin{pmatrix} g_0(z) \\ g_1(z) \end{pmatrix};$$

and we get

$$\begin{cases} g_0(z) + U(z^2) g_1(z) = A(z^2) + B(z^2)z \\ g_1(z) = (S_0^* f_1)(z^2) + (S_1^* f_1)(z^2)z. \end{cases}$$

Apply S_i^*, $i = 0, 1$

$$\Rightarrow \begin{cases} S_0^* g_0 + U S_0^* g_1 = A, \quad S_1^* g_0 + U S_1^* g_1 = B \\ S_0^* g_1 = S_0^* f_1, \quad S_1^* g_1 = S_1^* f_1 \\ \Rightarrow U = \dfrac{A - S_0^* g_0}{S_0^* f_1} \quad \text{and} \quad U = \dfrac{B - S_1^* g_0}{S_1^* f_1} \end{cases}$$

and continue.

$$S_0^* f_0 = A, \quad S_1^* f_0 = B$$

$$L S_0^* f_0 + S_0^* f_0 = C$$

$$L S_1^* f_0 + S_1^* f_0 = D$$

$$L = \frac{C - S_0^* f_1}{A} = \frac{D - S_1^* f_1}{B}$$

$$A(D - S_1^* f_1) = B(C - S_0^* f_1)$$

$$1 = A S_1^* f_1 - B S_0^* f_1$$

$$A_{\text{new}}^{(1)} = \begin{pmatrix} 1 & 0 \\ -L & 1 \end{pmatrix} \begin{pmatrix} A & B \\ C & D \end{pmatrix} = \begin{pmatrix} A & B \\ -LA+C & -LB+D \end{pmatrix} = \begin{pmatrix} A & B \\ S_0^* f_1 & S_1^* f_1 \end{pmatrix},$$

so

$$\begin{pmatrix} 1 & 0 \\ L & 1 \end{pmatrix} \begin{pmatrix} A & B \\ S_0^* f_1 & S_1^* f_1 \end{pmatrix} = \begin{pmatrix} A & B \\ C & D \end{pmatrix}.$$

7.3 Factorization Cases

In order to answer the main issue raised in Section 7.1, in the present section it will be helpful to discuss briefly the group SL_2 over the ring of L^∞ functions on $\mathbb{T}$. This discussion also serves to highlight the role of the C^*-algebra $\mathcal{O}_2$. The lemmas will extend from $N = 2$ to $N > 2$ with suitable modifications, as we outline in the subsequent section, but we find it useful to first present them in the case of $N = 2$ [JS14b].

In the infinite-dimensional group $SL_2(L^\infty(\mathbb{T}))$, consider elements $\mathcal{A}$ with factorization as in (7.7):

$$\mathcal{A} = \begin{pmatrix} A & B \\ C & D \end{pmatrix}$$

$$\mathcal{A} = \begin{pmatrix} 1 & 0 \\ L & 1 \end{pmatrix} \mathcal{A}^{(1)}, \quad L \in L^\infty(\mathbb{T}), \quad \mathcal{A}^{(1)} \in SL_2(L^\infty(\mathbb{T})) \qquad (7.7)$$

that gives optimal factorization:

$$\mathcal{A}^{(1)}(z^2) \begin{pmatrix} 1 \\ z \end{pmatrix} = \begin{pmatrix} f_0 \\ f_1 \end{pmatrix}$$

$$\begin{cases} A(z^2) + zB(z^2) = f_0(z) \\ C(z^2) + zD(z^2) = L(z^2)f_0(z) + f_1(z) \end{cases} \qquad \{S_i\}_{i=0} \in REP(\mathcal{O}_2, L^2(\mathbb{T})) \qquad (7.8)$$

$$\begin{cases} S_0^* f_0 = A, S_1^* f_0 = B \\ LS_0^* f_0 + S_0^* f_1 = C \\ LS_1^* f_0 + S_1^* f_1 = D \end{cases} \qquad (7.9)$$

$$\Longleftrightarrow \begin{cases} S_0^* f_1 = C - LA \\ S_1^* f_1 = D - LB \end{cases} \qquad (7.10)$$

$$\Rightarrow f_1 = (S_0 S_0^* + S_1 S_1^*) f_1 = S_0(C - LA) + S_1(D - LB) \qquad (7.11)$$

$$\begin{pmatrix} A & B \\ C & D \end{pmatrix} \longrightarrow \begin{pmatrix} A & B \\ S_0^* f_1 & S_1^* f_1 \end{pmatrix},$$

since S_i is isometric for $i = 1, 2$.

$$\|f_1\|^2 = \|C - LA\|^2 + \|D - LB\|^2 \quad \text{where } \|\cdot\| \text{ is the } L^2(\mathbb{T})\text{-norm.} \quad (7.12)$$

$$\langle u, v \rangle = \int_{\mathbb{T}} \overline{u}v \quad \text{with respect to Haar measure on } \mathbb{T}. \tag{7.13}$$

So for any function

$$\mathcal{A} = \begin{pmatrix} 1 & 0 \\ L & 1 \end{pmatrix} \mathcal{A}^{(1)} \tag{7.14}$$

we pick the one with f_1 attaching its minimum in (7.12)

$$\inf\{(7.12)|\text{factorization } (7.14) \text{ holds}\} \tag{7.15}$$

Calculating L on $\mathcal{A}$

$$L_M(\epsilon) = L + \epsilon M, \quad M \in L^\infty(\mathbb{T}).$$

$$\left. \frac{d}{d\epsilon} \right|_{\epsilon=0} (7.12) = 0 \quad \text{at a minimum.} \tag{7.16}$$

$$\Longleftrightarrow$$

$$\langle MA, C - LA \rangle + \langle C - LA, MA \rangle + \langle MB, D - LB \rangle + \langle D - LB, MB \rangle$$
$$= \text{Re}(\langle MA, C - LA \rangle + \langle MB, D - LB \rangle) = 0 \quad \forall M \in L^\infty(\mathbb{T}).$$

$$\overline{A}(C - LA) + \overline{B}(D - LB) = 0 \quad \text{pointwise a.e. on } \mathbb{T}. \tag{7.17}$$

Set $\det \mathcal{A} = 1$,

$$\|A\|^2 + \|B\|^2 > 0 \quad \text{a.e. on } \mathbb{T}.$$

So

$$L = \frac{\overline{A}C + \overline{B}D}{|A|^2 + |B|^2} \quad \text{pointwise a.e. } \mathbb{T}. \tag{7.18}$$

Solving for matrices $\mathcal{A}^{(1)}$ in (7.16), we get

$$\mathcal{A}^{(1)} = \begin{pmatrix} 1 & 0 \\ -L & 1 \end{pmatrix} \begin{pmatrix} A & B \\ C & D \end{pmatrix} = \begin{pmatrix} A & B \\ C - LA & D - LB \end{pmatrix}.$$

So

$$\mathcal{A} = \begin{pmatrix} 1 & 0 \\ L & 1 \end{pmatrix} \mathcal{A}^{(1)}.$$

With the above L in (7.18), we see that

$$\mathcal{A} = \begin{pmatrix} 1 & 0 \\ L & 1 \end{pmatrix} \mathcal{A}^{(1)}$$

is the *optimal* factorization with a lower matrix as a left-factor.

Corollary 7.2. *Given*

$$\begin{pmatrix} A & B \\ C & D \end{pmatrix} \in GL_2(L^\infty(\mathbb{T}));$$

then the optimal solution (7.18) *to the factorization problem*

$$\begin{pmatrix} A & B \\ C & D \end{pmatrix} = \begin{pmatrix} 1 & 0 \\ L & 1 \end{pmatrix} \begin{pmatrix} A & B \\ S_0^* f_1 & S_1^* f_1 \end{pmatrix} \tag{7.19}$$

has the matrix

$$\begin{pmatrix} A & B \\ S_0^* f_1 & S_1^* f_1 \end{pmatrix}$$

on the right-hand side in (7.19) *orthogonal, i.e.,*

$$\overline{A}(S_0^* f_1) + \overline{B}(S_1^* f_1) \equiv 0 \quad \text{on } \mathbb{T}. \tag{7.20}$$

Proof. When the function L in (7.18) is used in the computation of

$$\begin{pmatrix} A & B \\ S_0^* f_1 & S_1^* f_1 \end{pmatrix},$$

we see that for any $z \in \mathbb{T}$, $((S_0^* f_1)(z), (S_1^* f_1)(z))$ in $\mathbb{C}$ is in the orthogonal complement of $(A(z), B(z))$; indeed, with (7.18), we get

$$\overline{A}(S_0^* f_1) + \overline{B}(S_1^* f_1)$$
$$= \overline{A}\left(C - \frac{\overline{A}C + \overline{B}D}{|A|^2 + |B|^2} A \right) + \overline{B}\left(D - \frac{\overline{A}C + \overline{B}D}{|A|^2 + |B|^2} B \right)$$
$$= \overline{A}C + \overline{B}D - (\overline{A}C + \overline{B}D) \equiv 0;$$

i.e., a pointwise identity for functions on $\mathbb{T}$. $\square$

Corollary 7.3. *If* $A \in SU(L^\infty(\mathbb{T}))$ *(i.e., unitary) then* L *in* (7.18) *is 0 and so* $A = A^{(1)}$, *so the factorization follows as Equation* (7.19).

Proof.

$$A = \begin{pmatrix} A & B \\ C & D \end{pmatrix},$$

so unitary, which makes the rows orthogonal, $\overline{A}C + \overline{B}D = 0$, in the inner product on $\mathbb{C}^2$

$$\langle z, w \rangle = \overline{z_1} w_1 + \overline{z_2} w_2$$

and $|A|^2 + |B|^2 = 1$. $\square$

$$\mathcal{A}^{(1)} = \begin{pmatrix} A & B \\ S_0^* f_1 & S_1^* f_1 \end{pmatrix} \tag{7.21}$$

using

$$\begin{pmatrix} S_0^* g_0 & S_1^* g_0 \\ S_0^* f_1 & S_1^* f_1 \end{pmatrix}.$$

Note that we use this repeatedly on any $\mathcal{A}^{(1)} \in SL_2(L^\infty(\mathbb{T}))$, each time pick L such that the infimum in (7.12) is attained.

With the same argument, we factor matrix

$$\begin{pmatrix} 1 & U \\ 0 & 1 \end{pmatrix} \quad U \in L^\infty(\mathbb{T})$$

$$\mathcal{A} = \begin{pmatrix} 1 & U \\ 0 & 1 \end{pmatrix} \mathcal{A}^{(2)}, \quad \mathcal{A}^{(2)} \in SL_2(L^\infty(\mathbb{T})). \tag{7.22}$$

Set

$$\begin{pmatrix} g_0 \\ g_1 \end{pmatrix} = \mathcal{A}^{(2)}(z^2) \begin{pmatrix} 1 \\ z \end{pmatrix} \tag{7.23}$$

$$\begin{cases} A(z^2) + zB(z^2) = g_0 + U(z^2)g_1 \\ C(z^2) + zD(z^2) = g_1 \end{cases}$$

$$\begin{cases} A = S_0^* g_0 + U S_0^* g_1 \\ B = S_1^* g_0 + U S_1^* g_1 & \qquad S_0^* g_0 = A - UC, \quad S_1^* g_0 = B - UD \quad (7.24) \\ C = S_0^* g_1, D = S_1^* g_1 \end{cases}$$

$$g_0 = S_0 S_0^* g_0 + S_1 S_1^* g_1 = S_0(A - UC) + S_1(B - UD)$$

$$\|g_0\|^2 = \|A - UC\|^2 + \|B - UD\|^2 \tag{7.25}$$

such that (7.22) holds.

$$\begin{pmatrix} A & B \\ C & D \end{pmatrix} \longrightarrow \begin{pmatrix} A - UC & B - UD \\ C & D \end{pmatrix}, \quad S_0^* g_0 = A - UC, \quad S_1^* g_0 = B - UD.$$

Pick U such that

$$\overline{C}(A - UC) + \overline{D}(B - UD) = 0$$

$$U = \frac{\overline{C}A + \overline{D}B}{|C|^2 + |D|^2} \tag{7.26}$$

$$\mathcal{A}^{(2)} = \begin{pmatrix} S_0^* g_0 & S_1^* g_0 \\ C & D \end{pmatrix} \tag{7.27}$$

in (7.22).

If

$$\mathcal{A} = \begin{pmatrix} A & B \\ C & D \end{pmatrix} \in SL_2(L^\infty(\mathbb{T})),$$

then

$$U = \frac{\overline{C}A + \overline{D}B}{|C|^2 + |D|^2} = 0.$$

See (7.26), so the factorization

$$\mathcal{A} = \begin{pmatrix} 1 & U \\ 0 & 1 \end{pmatrix} \mathcal{A}^{(2)}$$

in (7.22) is then $U = 0 \Rightarrow \mathcal{A} = \mathcal{A}^{(2)}$. Then following factorization results:

$$\mathcal{A} = \left(\prod (\text{lower})(\text{upper}) \right) SL_2(L^\infty(\mathbb{T}))$$

$$\begin{pmatrix} A & B \\ C & D \end{pmatrix} \xrightarrow[\text{factor out lower matrix on the left}]{} \begin{pmatrix} A & B \\ S_0^* f_1 & S_1^* f_1 \end{pmatrix} \qquad (7.28)$$

$$\xrightarrow[\text{factor out upper matrix on the left}]{} \begin{pmatrix} S_0^* g_0 & S_1^* g_0 \\ S_0^* f_1 & S_1^* f_1 \end{pmatrix}.$$

Or equivalently,

$$\begin{pmatrix} A & B \\ C & D \end{pmatrix} = \begin{pmatrix} 1 & 0 \\ L & 1 \end{pmatrix} \begin{pmatrix} 1 & U \\ 0 & 1 \end{pmatrix} \begin{pmatrix} S_0^* g_0 & S_1^* g_0 \\ S_0^* f_1 & S_1^* f_1 \end{pmatrix}. \qquad (7.29)$$

Corollary 7.4. *Consider* $\mathcal{A} \in SL_2(L^\infty(\mathbb{T}))$*, and the factorization*

$$\mathcal{A} = \begin{pmatrix} 1 & 0 \\ L_1 & 1 \end{pmatrix} \begin{pmatrix} 1 & U_1 \\ 0 & 1 \end{pmatrix} \cdots \begin{pmatrix} 1 & 0 \\ L_p & 1 \end{pmatrix} \begin{pmatrix} 1 & U_p \\ 0 & 1 \end{pmatrix} \begin{pmatrix} S_0^* g_0 & S_1^* g_0 \\ S_0^* f_1 & S_1^* f_1 \end{pmatrix} \qquad (7.30)$$

resulting from an iteration of the algorithm from (7.29). Then the last factor in (7.30) is of diagonal form if and only if the following hold: There are functions $\varphi, \psi \in L^2(\mathbb{T})$ *such that*

$$g_0(z) = \varphi(z^2) \quad \text{and} \quad f_1(z) = z\psi(z^2); \qquad (7.31)$$

and, in this case, the last factor in (7.30) is as follows:

$$\begin{pmatrix} S_0^* g_0 & S_1^* g_0 \\ S_0^* f_1 & S_1^* f_1 \end{pmatrix} = \begin{pmatrix} \varphi & 0 \\ 0 & \psi \end{pmatrix}. \qquad (7.32)$$

Proof. This follows from (7.29), and the Cuntz relations:

$$S_i^* S_j = \delta_{i,j}, \quad \sum_i S_i S_i^* = I. \qquad (7.33)$$

$\square$

7.3.1 *Factorizations*

We fix a value of $N > 1$ (i.e., the given number of frequency bands), and we begin with the formula for a canonical system of N isometries S_i which define an associated representation of the Cuntz algebra $\mathcal{O}_N$. Said differently: The system of isometries $\{S_i\}$ satisfies the Cuntz relations with reference to the Hilbert space $L^2(\mathbb{T})$ where $\mathbb{T}$ is the circle group (one torus) with its normalized invariant Haar measure. When the value of N is fixed, then the multi-resolution filters will take the form of $N \times N$ matrix functions; the matrix entries might be polynomials, or, more generally, functions from $L^\infty(\mathbb{T})$. Hence, the questions about matrix factorization depend on the context. In the case of polynomial entries, we will make use of degree, but this is not available for the more general case of entries from the algebra $L^\infty(\mathbb{T})$. In every one of the settings, we develop factorization algorithms, and the particular representation of the Cuntz algebra will play an important role [JS14b].

Let the set of all orthogonal N-filters be denoted $\mathcal{OF}_N$ [JS10]. The standard representation of $\mathcal{O}_N$, which we will use in what follows, is given by the system of isometries $\{S_j\}$ as follows:

$$(S_j\varphi)(z) = f_j(z)\varphi(z^N). \tag{7.34}$$

Lemma 7.5. [JS10] *Let $N \in \mathbb{Z}_+$ be given and let $F = (f_j)_{j \in \mathbb{Z}_+}$ be a function system. Then $F \in \mathcal{OF}_N$ if and only if the operators S_j (7.34) satisfy*

$$S_j^* S_k = \delta_{j,k} I \tag{7.35}$$

$$\sum_{j \in \mathbb{Z}_N} S_j S_j^* = I, \tag{7.36}$$

where I denotes the identity operator in $\mathcal{H} = L^2(\mathbb{T})$.

We say that the isometries $\{S_j\}_{j \in \mathbb{Z}_N}$ define a representation of the Cuntz algebra $\mathcal{O}_N$, $(S_j) \in \text{Rep}(\mathcal{O}_N, L^2(\mathbb{T}))$.

Let $\mathcal{F}_2(N) := L^2(\mathbb{T}, \mathbb{C}^N) = \sum_0^{N-1} \oplus L^2(\mathbb{T})$ where the notation in the summation symbol means orthogonal direct sum with

$$\|F\|_2^2 = \sum_{j=0}^{N-1} \|f_j\|_{L^2(\mathbb{T})}^2 < \infty.$$

We will be making use of the special vector $b \in \mathcal{F}_2(N)$,

$$b(z) = \begin{bmatrix} 1 \\ z \\ z^2 \\ \vdots \\ z^{N-1} \end{bmatrix} ;$$

see [JS10].

Lemma 7.6. *[JS10] Let $N \in \mathbb{Z}_+$ be fixed, $N > 1$, and let $A = (A_{j,k})$ be an $N \times N$ matrix function with $A_{j,k} \in L^2(\mathbb{T})$. Then the following two conditions are equivalent:*

(i) *For $F = (f_j) \in \mathcal{F}_2(N)$, we have $F(z) = A(z^N)b(z)$.*
(ii) *$A_{i,j} = S_j^* f_i$ where the operators S_i are from the Cuntz relations (7.35, 7.36).*

See [JS10] for proof.

Theorem 7.7. *(Sweldens [SR91], [JS10]) Let $A \in SL_2(pol)$, then there are $l, p \in \mathbb{Z}_+$, $K \in \mathbb{C} \setminus \{0\}$, and polynomial functions $U_1, \ldots, U_p, L_1, \ldots, L_p$ such that*

$$A(z) = z^l \begin{pmatrix} K & 0 \\ 0 & K^{-1} \end{pmatrix} \begin{pmatrix} 1 & U_1(z) \\ 0 & 1 \end{pmatrix} \begin{pmatrix} 1 & 0 \\ L_1(z) & 1 \end{pmatrix} \cdots \begin{pmatrix} 1 & U_p(z) \\ 0 & 1 \end{pmatrix} \begin{pmatrix} 1 & 0 \\ L_p(z) & 1 \end{pmatrix}.$$
(7.37)

The filter algorithm corresponding to the matrix factorization in (7.37) is as follows: And in steps in Fig. 7.7:

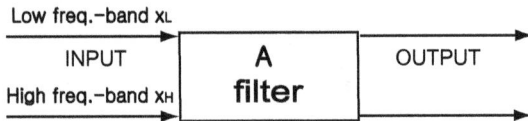

Fig. 7.7 Filters.

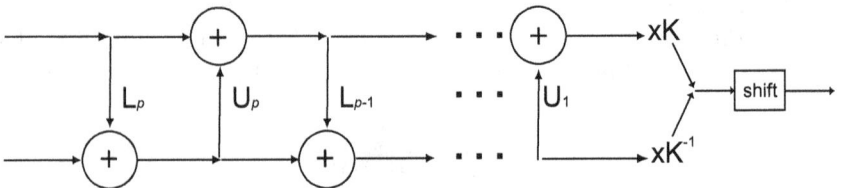

Remark 7.8. [JS10] Note that if

$$\begin{pmatrix} \alpha & \beta \\ \gamma & \delta \end{pmatrix} \in SL_2(\text{pol}),$$

then one of the two functions $\alpha(z)$ or $\delta(z)$ must be a monomial.

7.3.2 *The 2 × 2 case: polynomials*

From [JS10], to highlight the general ideas, we begin with some details worked out in the 2×2 case; see Equation (7.28).

To get finite algorithms, we should assume in the present subsection that the matrix entries are polynomials.

First note that from the setting in Theorem 7.7 we may assume that matrix entries have the form $f_H(z)$ but with $H \subset \{0, 1, 2, \ldots\}$, i.e., $f_H(z) = a_0 + a_1 z + \cdots$. This facilitates our use of the Euclidean algorithm [JS14b].

Specifically, if f and g are polynomials (i.e., $H \subset \{0, 1, 2, \ldots\}$) and if $\deg(g) \leq \deg(f)$, the Euclidean algorithm yields

$$f(z) = g(z)q(z) + r(z) \tag{7.38}$$

with $\deg(r) < \deg(g)$. We shall write

$$q = \text{quot}(g, f) \quad \text{and} \quad r = \text{rem}(g, f). \tag{7.39}$$

Since

$$\begin{pmatrix} K & 0 \\ 0 & K^{-1} \end{pmatrix} \begin{pmatrix} 1 & U \\ 0 & 1 \end{pmatrix} = \begin{pmatrix} 1 & K^2 U \\ 0 & 1 \end{pmatrix} \begin{pmatrix} K & 0 \\ 0 & K^{-1} \end{pmatrix}, \tag{7.40}$$

we may assume that the factor

$$\begin{pmatrix} K & 0 \\ 0 & K^{-1} \end{pmatrix}$$

from Equation (7.40) factorization occurs on the rightmost place.

$$F = U_N[b], \tag{7.41}$$

where U is a unitary matrix function, where

$$b = \begin{pmatrix} 1 \\ z \\ z^2 \\ \vdots \\ z^{N-1} \end{pmatrix}$$

and where $U_N[b](z) = U(z^N)b(z)$.

Let U represent a scalar-valued matrix entry in a matrix function. We now proceed to determine the polynomials $U_1(z)$, $L_1(z)$, ..., etc. inductively starting with

$$A = \begin{pmatrix} 1 & U \\ 0 & 1 \end{pmatrix} B,$$

where U and B are to be determined. Introducing (7.41), this reads

$$A(z^2) \begin{pmatrix} 1 \\ z \end{pmatrix} = \begin{pmatrix} 1 & U(z^2) \\ 0 & 1 \end{pmatrix} B(z^2) \begin{pmatrix} 1 \\ z \end{pmatrix} = \begin{pmatrix} 1 & U(z^2) \\ 0 & 1 \end{pmatrix} \begin{pmatrix} h(z) \\ k(z) \end{pmatrix}. \qquad (7.42)$$

But the matrix function

$$A = \begin{pmatrix} \alpha & \beta \\ \gamma & \delta \end{pmatrix}$$

is given and fixed, see Remark 7.8. Hence,

$$\gamma(z^2) + \delta(z^2)z = k(z) \qquad (7.43)$$

is also fixed. The two polynomials to be determined are U and h in (7.42). Carrying out the matrix product in (7.42) yields:

$$\alpha(z^2) + \beta(z^2)z = h(z) + U(z^2)k(z) = h_0(z) + h_1(z^2)z + U(z^2)\{\gamma(z^2) + \delta(z^2)z\},$$

where we used the orthogonal splitting

$$L^2(\mathbb{T}) = S_0 S_0^* L^2(\mathbb{T}) \oplus S_1 S_1^* L^2(\mathbb{T}) \qquad (7.44)$$

from Lemma 7.5. Similarly, from (7.43), we get

$$\gamma(z^2) + \delta(z^2)z = k_0(z^2) + k_1(z^2)z;$$

and therefore $\gamma = k_0$ and $\delta = k_1$, by Lemma 7.6.

Collecting terms and using the orthogonal splitting (7.44), we arrive at the following system of polynomial equations:

$$\begin{cases} \alpha = h_0 + U\gamma \\ \beta = h_1 + U\delta; \end{cases} \qquad (7.45)$$

or more precisely,

$$\begin{cases} \alpha(z) = h_0(z) + U(z)\gamma(z) \\ \beta(z) = h_1(z) + U(z)\delta(z). \end{cases}$$

It follows that the two functions U and h may be determined from the Euclidean algorithm. With (7.40), we get

$$\begin{cases} U = \text{quot}(\gamma, \alpha) \\ h_0 = \text{rem}(\gamma, \alpha) \\ h_1 = \text{rem}(\delta, \beta). \end{cases} \quad (7.46)$$

Remark 7.9. [JS10] The relevance of the determinant condition we have from Theorem 7.7 is as follows:

$$\det A = \alpha\delta - \beta\gamma \equiv 1.$$

Substitution of (7.45) into this yields

$$h_0\delta - h_1\gamma \equiv 1.$$

Solutions to (7.45) are possible because the two polynomials $\delta(z)$ and $\gamma(z)$ are mutually prime. The derived matrix

$$\begin{pmatrix} h_0 & h_1 \\ \gamma & \delta \end{pmatrix}$$

is obtained from A via a row operation in the ring of polynomials.

For the inductive step, it is important to note as follows:

$$\deg(h_0) < \deg(\gamma) \quad \text{and} \quad \deg(h_1) < \deg(\delta). \quad (7.47)$$

The next step, continuing from (7.42), is the determination of a matrix function C and three polynomials p, q, and L such that

$$\begin{pmatrix} 1 & -U \\ 0 & 1 \end{pmatrix} A = \begin{pmatrix} 1 & 0 \\ L & 1 \end{pmatrix} C \quad (7.48)$$

and

$$\begin{pmatrix} 1 & -U(z^2) \\ 0 & 1 \end{pmatrix} A(z^2) \begin{pmatrix} 1 \\ z \end{pmatrix} = \begin{pmatrix} 1 & 0 \\ L(z^2) & 1 \end{pmatrix} \begin{pmatrix} p(z) \\ q(z) \end{pmatrix}. \quad (7.49)$$

Here,

$$\begin{pmatrix} p \\ q \end{pmatrix} = C(z^2) \begin{pmatrix} 1 \\ z \end{pmatrix}.$$

The reader will note that in this step everything is as before, with the only difference that now

$$\begin{pmatrix} 1 & 0 \\ L & 1 \end{pmatrix}$$

is lower diagonal in contrast with

$$\begin{pmatrix} 1 & U \\ 0 & 1 \end{pmatrix}$$

in the previous step.

This time, the determination of the polynomial p in (7.49) is automatic. With

$$p(z) = p_0(z^2) + zp_1(z^2)$$

(see (7.44)), we get the following system:

$$\begin{cases} p_0 = \alpha - U\gamma = h_0 \\ p_1 = \beta - U\delta = h_1; \quad \text{and} \end{cases}$$

$$\begin{cases} \gamma = L(\alpha - U\gamma) + q_0 = Lh_0 + q_0 \\ \delta = L(\beta - U\delta) + q_1 = Lh_1 + q_1. \end{cases}$$

So the determination of $L(z)$ and $q(z) = q_0(z^2) + zq_1(z^2)$ may be done with Euclid:

$$\begin{cases} L = \text{quot}(\alpha - U\gamma, \gamma) = \text{quot}(h_0, \gamma) \\ q_0 = \text{rem}(\alpha - U\gamma, \gamma) = \text{rem}(h_0, \gamma) \\ q_1 = \text{rem}(\beta - U\delta, \delta) = \text{rem}(h_1, \delta). \end{cases} \tag{7.50}$$

Combining the two steps, the comparison of degrees is as follows:

$$\begin{cases} \deg(q_0) < \deg(h_0) < \deg(\gamma) \\ \deg(q_1) < \deg(h_1) < \deg(\delta). \end{cases} \tag{7.51}$$

Two conclusions now follow:

(1) the procedure may continue by recursion;
(2) the procedure must terminate.

Remark 7.10. In order to start the algorithm in (7.46) with direct reference to Euclid, we must have

$$\deg(\gamma) \leq \deg(\alpha) \tag{7.52}$$

where

$$A = \begin{pmatrix} \alpha & \beta \\ \gamma & \delta \end{pmatrix}$$

is the initial 2×2 matrix function.

Now, suppose (7.52), i.e., that

$$\deg(\gamma) \le \deg(\alpha).$$

Then determine a polynomial L such that

$$\deg(\gamma - L\alpha) \le \deg(\alpha). \tag{7.53}$$

We may then start the procedure (7.46) on the matrix function

$$\begin{pmatrix} \alpha & \beta \\ \gamma - L\alpha & \delta \end{pmatrix} = \begin{pmatrix} 1 & 0 \\ -L & 1 \end{pmatrix} A.$$

If a polynomial U and a matrix function B are then found for

$$\begin{pmatrix} \alpha & \beta \\ \gamma - L\alpha & \delta \end{pmatrix},$$

then the factorization

$$A = \begin{pmatrix} 1 & 0 \\ L & 1 \end{pmatrix} \begin{pmatrix} 1 & U \\ 0 & 1 \end{pmatrix} B$$

holds; and the recursion will then work as outlined.

In the following, starting with a matrix function A, we will always assume that the degrees of the polynomials $(A_{i,j})_{i,j \in \mathbb{Z}_N}$ have been adjusted this way, so the direct Euclidean algorithm can be applied [JS14b].

7.3.3 *The* 3×3 *case*

The thrust of this section is the assertion that Theorem 7.7 holds with small modifications in the 3×3 case.

7.3.3.1 *Comments*

In the definition of $A \in SL_3(\text{pol})$, it is understood that $A(z)$ has $\det A(z) \equiv 1$ and that the entries of the inverse matrix $A(z)^{-1}$ are again polynomials [JS14b].

Note that if $L, M, U,$ and V are polynomials, then the four matrices

$$\begin{pmatrix} 1 & 0 & 0 \\ L & 1 & 0 \\ 0 & M & 1 \end{pmatrix}, \begin{pmatrix} 1 & 0 & 0 \\ 0 & 1 & 0 \\ L & 0 & 1 \end{pmatrix}, \begin{pmatrix} 1 & U & 0 \\ 0 & 1 & V \\ 0 & 0 & 1 \end{pmatrix}, \text{ and } \begin{pmatrix} 1 & 0 & U \\ 0 & 1 & 0 \\ 0 & 0 & 1 \end{pmatrix} \tag{7.54}$$

are in $SL_3(\text{pol})$ since

$$\begin{pmatrix} 1 & 0 & 0 \\ L & 1 & 0 \\ 0 & M & 1 \end{pmatrix}^{-1} = \begin{pmatrix} 1 & 0 & 0 \\ -L & 1 & 0 \\ LM & -M & 1 \end{pmatrix} \quad \text{and} \quad (7.55)$$

$$\begin{pmatrix} 1 & U & 0 \\ 0 & 1 & V \\ 0 & 0 & 1 \end{pmatrix}^{-1} = \begin{pmatrix} 1 & -U & UV \\ 0 & 1 & -V \\ 0 & 0 & 1 \end{pmatrix}. \quad (7.56)$$

Theorem 7.11. *Let $A \in SL_3(\text{pol})$; then the conclusion in Theorem 7.7 carries over with the modification that the alternating upper and lower triangular matrix functions now have the form (7.54) or (7.55)–(7.56) where the functions L_j, M_j, U_j, and V_j, $j = 1, 2, \ldots$ are polynomials.*

7.3.4 *The $N \times N$ case*

Theorem 7.12. *[JS14b] Let $N \in \mathbb{Z}_+$, $N > 1$, be given and fixed. Let $A \in SL_N(\text{pol})$; then the conclusions in Theorem 7.7 carry over with the modification that the alternative factors in the product are upper and lower triangular matrix functions in $SL_N(\text{pol})$. We may take the lower triangular matrix factors $\mathcal{L} = (L_{i,j})_{i,j \in \mathbb{Z}_N}$ of the form*

$$\begin{pmatrix} 1 & 0 & 0 & 0 & 0 & 0 & 0 & 0 \\ 0 & 1 & 0 & 0 & 0 & 0 & 0 & 0 \\ L_p & 0 & 1 & 0 & 0 & 0 & 0 & 0 \\ 0 & L_{p+1} & 0 & 1 & 0 & 0 & 0 & 0 \\ 0 & 0 & . & 0 & 1 & 0 & 0 & 0 \\ 0 & 0 & 0 & . & 0 & 1 & 0 & 0 \\ 0 & 0 & 0 & 0 & . & 0 & 1 & 0 \\ 0 & 0 & 0 & 0 & 0 & L_{N-1} & 0 & 1 \end{pmatrix},$$

polynomial entries of the form

$$\begin{cases} L_{i,i} \equiv 1, \\ L_{i,j}(z) = \delta_{i-j,p} L_i(z); \end{cases} \quad (7.57)$$

and the upper triangular factors of the form $\mathcal{U} = (U_{i,j})_{i,j \in \mathbb{Z}_N}$ with

$$\begin{cases} U_{i,i} \equiv 1, \\ L_{i,j}(z) = \delta_{i-j,p} U_i(z). \end{cases} \quad (7.58)$$

Proof. Notation. Let $U_1, \ldots, U_N, L_1, \ldots, L_N$ be polynomials and set

$$
\mathcal{U}_N(U) = \begin{pmatrix}
1 & U_1 & 0 & 0 & 0 & 0 & 0 \\
0 & 1 & U_2 & 0 & 0 & 0 & 0 \\
0 & 0 & 1 & . & 0 & 0 & 0 \\
0 & 0 & 0 & 1 & . & 0 & 0 \\
0 & 0 & 0 & 0 & 1 & . & 0 \\
0 & 0 & 0 & 0 & 0 & 1 & U_{N-1} \\
0 & 0 & 0 & 0 & 0 & 0 & 1
\end{pmatrix} \tag{7.59}
$$

$$
\mathcal{L}_N(L) = \begin{pmatrix}
1 & 0 & 0 & 0 & 0 & 0 & 0 \\
L_1 & 1 & 0 & 0 & 0 & 0 & 0 \\
0 & L_2 & 1 & 0 & 0 & 0 & 0 \\
0 & 0 & . & 1 & 0 & 0 & 0 \\
0 & 0 & 0 & . & 1 & 0 & 0 \\
0 & 0 & 0 & 0 & . & 1 & 0 \\
0 & 0 & 0 & 0 & 0 & L_{N-1} & 1
\end{pmatrix}. \tag{7.60}
$$

Note that both are in $SL_N(\text{pol})$; and we have

$$
\mathcal{U}_N(U)^{-1} = \mathcal{U}_N(-U) \quad \text{and}
$$
$$
\mathcal{L}_N(L)^{-1} = \mathcal{L}_N(-L).
$$

Step 1: Starting with $A = (A_{i,j}) \in SL_N(\text{pol})$. Then left-multiply with a suitably chosen $\mathcal{U}_N(-U)$ such that the degrees in the first column of $\mathcal{U}_N(-U)A$ decrease, i.e.,

$$
\deg(A_{0,0}) \le \deg(A_{1,0} - u_2 A_{1,0}) \le \cdots \deg(A_{N-1,0}). \tag{7.61}
$$

In the following, we shall use the same letter A for the modified matrix function.

Step 2: Determine a system of polynomials $L_1, \ldots, L_{N-1}$ and a polynomial vector function

$$
\begin{bmatrix} f_0 \\ f_1 \\ \ldots \\ f_{N-1} \end{bmatrix}
$$

such that

$$
A_N \begin{bmatrix} 1 \\ z \\ z^2 \\ \ldots \\ z^{N-1} \end{bmatrix} = \mathcal{L}_N(L)_N \begin{bmatrix} f_0 \\ f_1 \\ \ldots \\ f_{N-1} \end{bmatrix}, \tag{7.62}
$$

or equivalently

$$\sum_{j=0}^{N-1} A_{i,j}(z^N)z^j = \begin{cases} f_0(z) & \text{if } i = 0 \\ L_i(z^N)f_{i-1}(z) + f_i(z) & \text{if } i > 0. \end{cases}$$

Step 3: Apply the operators S_j and S_j^* from (6.21) to both sides in (7.62). First, (7.62) takes the form

$$\sum_{j=0}^{N-1} S_j A_{i,j} = \begin{cases} f_0 & \text{if } i = 0 \\ S_{f_{i-1}} L_i + f_i & \text{if } i > 0. \end{cases}$$

For $i = 1$, we get

$$A_{1,j} = L_1 A_{0,j} + k_j \quad \text{where } k_j = S_j^* f_1. \tag{7.63}$$

By (7.61) and the assumptions on the matrix functions, we note that the system (7.63) may now be solved with the Euclidean algorithm:

$$\begin{cases} L_1 = \text{quot}(A_{0,j}, A_{1,j}) \\ k_j = \text{rem}(A_{0,j}, A_{1,j}) \end{cases} \tag{7.64}$$

with the same polynomial L_1 for $j = 0, 1, \ldots, N - 1$.

For the polynomial function f_1, we then have

$$f_1 = \sum_{j=0}^{N-1} S_j k_j; \tag{7.65}$$

i.e.,

$$f_1(z) = k_0(z^N) + k_1(z^N)z + \cdots + k_{N-1}(z^{N-1})z^{N-1}.$$

The process now continues recursively until all the functions $L_1, L_2, \ldots, f_1, f_2, \ldots$ have been determined.

Step 4: The formula (7.62) translates into a matrix factorization as follows: With L and F determined in (7.62), we get

$$A = \mathcal{L}_N(L)B \tag{7.66}$$

as a simple matrix product taking $B = (B_{i,j})$ and

$$B_{i,j} = S_j^* f_i, \tag{7.67}$$

where we used Lemmas 7.5 and 7.6.

Step 5: The process now continues with the polynomial matrix function from (7.66) and (7.67). We determine polynomials $U_1, \ldots, U_{N-1}$ and a third matrix function

$$C = (C(z)) = (C_{i,j}(z)) \quad \text{such that} \quad B = \mathcal{U}_N(U)C.$$

Step 6: At each step of the process, we alternate L and U; and at each step, the degrees of the matrix functions are decreased. Hence, the recursion must terminate as stated in Theorem 7.12.   □

7.3.5 $L^\infty(\mathbb{T})$-matrix entries

While the case $N = 2$ is motivated by application to the high-pass vs. low-pass filters. The conclusions for $N = 2$ carry over to $N > 2$ cases. To see this, we first define the Cuntz algebra $\mathcal{O}_N$, in general the relations are

$$S_i^* S_j = \delta_{i,j} I, \quad \sum_i S_i S_i^* = I, \tag{7.68}$$

when the elements $(S_i)_{i=0}^{N-1}$ are given to be symmetric.

Each case (7.68) has many representations; e.g., if $(m_i(z))_{i=0}^{N-1}$, $z \in \mathbb{T}$, is a system of filters corresponding to N frequency bands, we may obtain a representation of $\mathcal{O}_N$ acting on the Hilbert space $L^2(\mathbb{T})$ as follows:

$$(S_i \psi)(z) = m_i(z)\psi(z^N), \quad \forall z \in \mathbb{T}, \quad \psi \in L^2(\mathbb{T}). \tag{7.69}$$

For $i \in \{0, 1, \ldots, N-1\}$, the adjoint operator of S_i in (7.69) is

$$(S_i^* \psi)(z) = \frac{1}{N} \sum_{w^N = z} \overline{m_i(z)}\psi(z^N), \quad z \in \mathbb{T}. \tag{7.70}$$

A direct verification shows that the Cuntz relation (7.68) is satisfied for the operators $(S_i)_{i=0}^{N-1}$ in (7.69) if and only if the system $(m_i)_{i=0}^{N-1}$ is a multiband filter covering the N frequency bands [JS14b].

The simplest example of the representation in (7.69) is the case where $m_i(z) = z^i$, $i = 0, 1, \ldots, N-1$; and so

$$(S_i \psi)(z) = z^i \psi(z^N), \quad i = 0, 1, \ldots, N-1, \quad z \in \mathbb{T}, \quad \psi \in L^2(\mathbb{T}). \tag{7.71}$$

Theorem 7.13. *Let* $g = (g_{ij})_{i,j=0}^{N-1} \in SL_N(L^\infty(\mathbb{T}))$, *i.e.,* $g_{ij}(\cdot) \in L^\infty(\mathbb{T}))$, *and*

$$\det g(\cdot) \equiv 1 \quad \text{on } \mathbb{T} \tag{7.72}$$

then for every factorization

$$g(z) = \begin{pmatrix} 1 & 0 & 0 & \cdots & 0 \\ L_1(z) & 1 & 0 & \cdots & 0 \\ L_2(z) & 0 & 1 & \ddots & 0 \\ \vdots & \vdots & \vdots & \ddots & \vdots \\ \vdots & \vdots & \vdots & \ddots & \vdots \\ L_{N-1}(z) & 0 & \cdots & 0 & 1 \end{pmatrix} g^{(\text{new})}(z) \quad \text{(matrix product)} \tag{7.73}$$

there is a unique $f_i \in L^\infty(\mathbb{T})$ such that

$$g_{0,j}^{(\text{new})}(z) = g_{0,j}(z), \quad \text{and} \tag{7.74}$$

$$g_{i,j}^{(\text{new})}(z) = S_j^* f_i, \quad \text{for } i = 1, 2, \ldots, N-1, \tag{7.75}$$

where $\{S_j\}_{j=0}^{N-1}$ is the system of Cuntz isometries from (7.71).

Proof. With the arguments above, in the space $\mathcal{O}_N$ of $N = 2$, we now get the matrix system

$$g^{(\text{new})}(z^N) \begin{pmatrix} 1 \\ z \\ z^2 \\ \vdots \\ z^{N-1} \end{pmatrix} = \begin{pmatrix} f_0(z) \\ f_1(z) \\ f_2(z) \\ \vdots \\ f_{N-1}(z) \end{pmatrix}, \tag{7.76}$$

$$\begin{cases} S_j^* f_0 = g_{0,j}, \\ L_i g_{0,j} + S_j^* f_i = g_{i,j}, \end{cases} \tag{7.77}$$

and

$$f_i = \sum_{j=0}^{N-1} S_j S_j^* f_i = \sum_{j=0}^{N-1} S_j(g_{i,j} - L_i g_{0,j}) \tag{7.78}$$

for $i = 1, 2, \ldots, N-1$, which is the desired conclusion. $\square$

7.3.6 Optimal factorization in the case of $SL_N(L^\infty(\mathbb{T}))$

Fix $N > 2$, and consider the usual inner product in $\mathbb{C}^N$,

$$\langle z, w \rangle := \sum_{j=0}^{N-1} \overline{z_j} w_j, \tag{7.79}$$

defined for all $z = (z_0, \ldots, z_{N-1})$, and $w = (w_0, \ldots, w_{N-1})$.

For $g = (g_{ij}(z))_{i,j=0}^{N-1} \in SL_N(L^\infty(\mathbb{T}))$, set

$$\tilde{g}_0(z) = (g_{0j}(z))_{j=0}^{N-1}, \quad \text{a.e.,}$$

the first row in the matrix function $\mathbb{T} \ni Z \mapsto (g(z)) \in SL_N(L^\infty(\mathbb{T}))$. Let $P(z) = P^{(g)}(z)$ denote the projection of $\mathbb{C}^N$ onto the one-dimensional subspace generated by $\tilde{g}_0(z) \in \mathbb{C}^N$.

Note that $(P(z))_{z \in \mathbb{T}}$ is a field of orthogonal rank-2 projections in $\mathbb{C}^N$. Setting

$$\|\tilde{g}_0(z)\|_2^2 = \sum_{j=0}^{N-1} |g_{0,j}(z)|^2, \tag{7.80}$$

we have,

$$P(z)\xi = \sum_{j=0}^{N-1} \frac{\overline{g_{0,j}(z)\xi_j}}{\|\tilde{g}_0(z)\|_2^2} \, g_{0,j}(z) \quad \text{for all } \xi = (\xi_0, \ldots, \xi_{N-1}) \in \mathbb{C}^N; \tag{7.81}$$

and set

$$\tilde{g}_j^{(\text{new})}(z) = \tilde{g}_0(z) - P(z)\tilde{g}_j(z). \tag{7.82}$$

Corollary 7.14.

(i) *For the factorization (7.73) in Theorem 7.13, the optimal choice is that given by the matrix factor $f^{(\text{new})}$ having as rows the vector fields $\tilde{g}_i^{(\text{new})}(z)$ specified in (7.82).*

(ii) *With the resolution of row-vector fields,*

$$\tilde{g}_j^{(\text{new})}(z) = (S_0^* f_i, S_1^* f_i, \ldots, S_{N-1}^* f_i) \tag{7.83}$$

from (7.75), the optimal solution is attained; and it is the unique minimizer for the following system of optimization problems:

$$\min_{f_i \in L^2(\mathbb{T})} \|f_i\|_{L^2(\mathbb{T})}^2, \quad 1 \leq i < N, \tag{7.84}$$

where each choice $(f_i)_{i=1}^{N-1}$ yields a matrix factor $\mathcal{A}^{(\text{new})}$ via (7.83).

Proof. The proof of the conclusions in (i)–(ii) in the corollary follows from the arguments in the proof of Theorem 7.13. $\qquad\square$

Appendix A

Hilbert Space Basics

As some of the readers may be new to the concept of Hilbert space, we include some basics from [JT21] for reference.

Definition A.1 (Inner product). Let X be a vector space over $\mathbb{C}$. An inner product on X is a function $\langle \cdot, \cdot \rangle : X \times X \to \mathbb{C}$ such that for all $x, y \in \mathcal{H}$, and $c \in \mathbb{C}$, we have

(a) $\langle x, \cdot \rangle : X \to \mathbb{C}$ is linear (linearity);
(b) $\langle x, y \rangle = \overline{\langle y, x \rangle}$ (conjugation);
(c) $\langle x, x \rangle \geq 0$; and $\langle x, x \rangle = 0$ implies $x = 0$ (positivity).

Definition A.2. Let $\mathcal{H}$ be a Hilbert space. A family of vectors $\{u_\alpha\}$ in $\mathcal{H}$ is said to be an *orthonormal basis* if

(1) $\langle u_\alpha, u_\beta \rangle_{\mathcal{H}} = \delta_{\alpha\beta}$, and
(2) $\overline{span} \{u_\alpha\} = \mathcal{H}$. (Here "$\overline{span}$" means "closure of the linear span.")

Corollary A.3. *For all $T \in \mathcal{B}(\mathcal{H})$, there exists a unique operator $T^* \in \mathcal{B}(\mathcal{H})$, called the adjoint of T, such that*
$$\langle x, Ty \rangle = \langle T^*x, y \rangle, \ \forall x, y \in \mathcal{H};$$
and $\|T^*\| = \|T\|$.

Definition A.4. Let $T \in \mathcal{B}(\mathcal{H})$.

(a) T is *normal* if $TT^* = T^*T$;
(b) T is *self-adjoint* if $T = T^*$;
(c) T is *unitary* if $T^*T = TT^* = I_{\mathcal{H}}$ ($=$ the identity operator);
(d) T is a (self-adjoint) *projection* if $T = T^* = T^2$.

Definition A.5. Let $\mathcal{H}$ be a Hilbert space with inner product $\langle \cdot, \cdot \rangle$. We denote by "bra" the vectors $\langle x|$ and by "ket", the vectors $|y\rangle$, for $x, y \in \mathcal{H}$.

With Dirac's notation, our first observation is the following lemma.

Lemma A.6. *Let $v \in \mathcal{H}$ be a unit vector. The operator $x \mapsto \langle v, x \rangle v$ can be written as $P_v = |v\rangle\langle v|$, i.e., a "ket-bra" vector (Definition A.5). And P_v is a rank-one self-adjoint projection.*

Corollary A.7. *Let $F = \operatorname{span}\{v_i\}$ with $\{v_i\}$ a finite set of orthonormal vectors in $\mathcal{H}$, then*

$$P_F := \sum_{v_i \in F} |v_i\rangle\langle v_i|$$

is the self-adjoint projection onto F.

Lemma A.8. *Let $\{u_\alpha\}_{\alpha \in J}$ be an ONB in $\mathcal{H}$, then we may write*

$$I_{\mathcal{H}} = \sum_{\alpha \in J} |u_\alpha\rangle\langle u_\alpha|.$$

Appendix B

Factorization of Matrices, Algorithms, and Wavelets*

As noted in Chapter 1, there are two sides to wavelet multiresolution theory; one is (i) the algorithmic part, and the other is (ii) a suitable choice of realization. In what follows, we shall demonstrate how part (i) can be realized in the form of filters, also called *filter banks*. By this, we refer to the kind of multiple frequency filter systems which are used in the architecture of signal/image processing algorithms. For (ii), we make use of wavelet realizations in the standard Hilbert $L^2(\mathbb{R})$. However, later we will show that $L^2(\mathbb{R})$ is far from the only choice for wavelet systems; e.g., we show that L^2 spaces of fractal measures are versatile Hilbert spaces for (ii), as are solenoid constructions commonly used in dynamics. Among the fractal measures that allow wavelet systems, we include *iterated function system* (IFS) measures, and, in particular, the maximal entropy measures associated with Julia sets in complex dynamics [Jor03]. The novice might find the following references helpful: [Bra06, BY06, Jor05, Law99]. This chapter is inspired by [Jor03].

What is a wavelet? In its simplest form, it is a function ψ on the real line $\mathbb{R}$, such that the doubly indexed family $\left\{2^{n/2}\psi\left(2^n x - k\right)\right\}_{n,k\in\mathbb{Z}}$ provides a basis for all the functions in a suitable space such as $L^2(\mathbb{R})$. Since $L^2(\mathbb{R})$ comes with a norm and inner product, it is natural to ask that the basis functions be normalized and mutually orthogonal (but many useful wavelets are not orthogonal). The analog-to-digital problem from

---

*With illustrations by Brian Treadway.

signal processing concerns the correspondence

$$f(x) \longleftrightarrow c_{n,k} \tag{B.1}$$

for the representation

$$f(x) = \sum_{n \in \mathbb{Z}} \sum_{k \in \mathbb{Z}} c_{n,k} 2^{n/2} \psi(2^n x - k). \tag{B.2}$$

We will be working primarily with the Hilbert space $L^2(\mathbb{R})$, and we allow complex-valued functions. Hence, the inner product $\langle f \mid g \rangle = \int \overline{f(x)} g(x)\, dx$ has a complex conjugate on the first factor in the product under the integral sign. If f represents a signal in analog form, the wavelet coefficients $c_{n,k}$ offer a digital representation of the signal, and the correspondence between the two sides in (B.1) is a new form of the analysis/synthesis problem — quite analogous to Fourier's analysis/synthesis problem of classical mathematics. One reason for the success of wavelets is the fact that the algorithms for the problem (B.1) are faster than the classical ones in the context of Fourier [Jor03].

Other efficient algorithms include the fast Fourier transform [Wic94] applied to digitized signals. It is based on dyadic scaling, a feature it shares with the hierarchical wavelet algorithms. But the latter have further advantages related to localization (see, e.g., [Mal09] and Section 11 of [Mey00b]. At discontinuities or sharp spikes, the edge effects for wavelet algorithms in the analog-to-digital problem are moderate, and they do not build up as in the Gibbs phenomenon with Fourier-based tools. An intrinsic feature of the subdivision scheme of wavelets is that the edge effects are concentrated in a few large wavelet coefficients, allowing us to neglect the rest; the asymptotic gain compared to Fourier series is exponential. A color picture is usually made up of a few homogeneous chunks of different colors or shades, one separated from the others by edges. It is for this reason that the wavelet algorithm is so much more efficient in digitizing graphics than are the alternative Fourier-based algorithms. Similarly, for signals, wavelet analysis breaks up $f(x)$ into its frequency components with each component in a resolution in time that is matched to its scale.

In what follows, we will want readers to relate our explanation to the two figures. So it will be good to add citations to them, one scale of resolution, and the other detail spaces (Figures B.1 and B.2)

A Hilbert space $\mathcal{H}$ is a real or complex inner product space that is also a complete metric space with respect to the distance function induced by the inner product.

Definition B.1. A Hilbert space $\mathcal{H}$ is a vector space with an inner product $\langle f, g \rangle$ such that the norm is defined by $\|f\| = \sqrt{\langle f, f \rangle}$ [Jor05], [BJ02a].

Hilbert space: Figures B.1 and B.2 are beautiful illustrations of the usefulness of Hilbert space in understanding more traditional multiresolutions.

B.1 Resolution and Detail

In our approach to wavelet mathematics, we are using the word *"multiresolution"* in a way that (deliberately) allows for multiple meanings, e.g., referring to scales of subspaces, recursive wavelet algorithms, resolution of digital images, from cameras, or from fingerprint readers, or from *frequency bands* in the transmission of wireless signals. For details on the precise meaning of and hands-on use of some of this, see Chapters 3 and 4. It is our hope that this multitude of occurrences of *multiresolutions* in pure and applied mathematics will in fact make it easier for students to understand some of the abstract, *Hilbert space theoretic*, variants of the basic concept of *scales* in "multiresolutions." Many such scales make references to nested families of closed subspaces in Hilbert space (see Figures B.1 and B.2 and Section B.4.) Moreover, these *scales of subspaces* in turn make direct connection to our representations of *wavelet functions* (see, e.g., (B.2) and Figure B.3). For matrix scaling, the same idea from (B.2) also takes many other more refined forms. Equations (B.13) and (B.14) represent a case in point. Equation (B.13) is called the *scaling identity* for the (wavelet) *father function* ψ (or scaling function), and (B.14) the scaling formula for the *mother functions*, ψ_i. When the operations of scaling and translations are applied to the *mother functions*, one then arrives at a wavelet basis, but the "quality" of such a wavelet basis representation depends on the selection of the coefficients in (B.13) and (B.14), the so-called wavelet coefficients; a topic discussed in detail in Chapters 2, 3, and 4.

In Chapter 6, we illustrate *multiresolution* for fractals. The reason many fractals admit *multiresolutions* is the inherent *similarity up to scale*, with suitable choices of scale; an idea which is inherent in the definition of a fractal. A key step in useful analyses of fractals will therefore be to make precise *similarity up to scale*.

The basic idea is that the wavelet basis functions from (B.4) will lie in the intermediate detail subspaces from Figures B.1 and B.2 As it turns out, our use of the notion of "multiresolution" also makes direct connection to what is commonly understood to be scales of resolution in the representation of *digital images*, see Section 2.3. There are many reasons for selecting Hilbert space for the representation of multiresolutions. For one, Hilbert space offers a useful tool in design of algorithms for computation of the coefficients in wavelet representations, or in digital image encoding. Indeed, once a suitable Hilbert space has been chosen, then the representation of a finer resolution will demand a bigger subspace of the ambient Hilbert space, while a coarser resolution only requires a smaller subspace. A key feature is then that the "intermediate detail" in a choice of resolution scales will correspond to the gap between pairs of resolution subspaces. Again, we refer to the visual representation of this idea from Figures B.1 and B.2; as well as its variants from applications presented in later chapters. For example, the representation of analysis on fractals, see Figures C.2 and C.1, again serves to illustrate the notion of "multiresolution." The same basic concept underlies representations of digital images; then the corresponding *resolution/detail data* can be represented in multiresolution form as illustrated in Figures C.1 and C.2, from Chapter 8; Section 6.5.

Figures B.1 and B.2 describe the resolution and detail pictorially. Mathematically, a *multiresolution* is a telescoping family of closed subspaces in some Hilbert space, doubly infinite, but generated by a single operator U and a single subspace $\mathcal{V}_0$. The operator U represents some scaling, and $\mathcal{V}_0$, some fixed resolution, e.g., step functions of step size one; the larger spaces in the family represent a finer resolution, and the smaller spaces, coarser resolutions. When applied to signal processing or to optics, each resolution space is assigned a band of frequencies. Matrix factorizations have a long history as a tool for designing fast algorithms. Factorizations have been used from the beginning in classical algorithms of signal processing, and, more recently, in wavelet subdivision schemes. Since algorithms for quantum information are also based on factorization of unitary matrices, it is not surprising that the subdivision-wavelet algorithms have proved to adapt

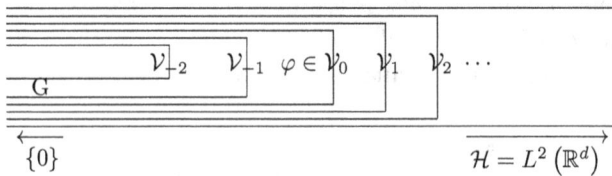

Fig. B.1   The subspaces of a resolution.

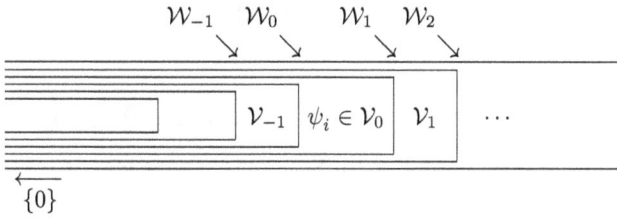

Fig. B.2   Incremental detail.

especially well to the realm of qubits in quantum theory: see, e.g., [PBK07] in the case of the Grover search algorithm.

The wavelet algorithms can be cast geometrically in terms of subspaces in Hilbert space which describe a scale of resolutions of some signal, or some picture. They are tailor-made for an algorithmic approach, which is based upon unitary matrices or upon functions with values in the unitary matrices. Wavelet analysis takes place in some Hilbert space $\mathcal{H}$ of functions on $\mathbb{R}^d$, e.g., $\mathcal{H} = L^2\left(\mathbb{R}^d\right)$. An indexed family of closed subspaces $\{\mathcal{V}_n\}_{-\infty < n < \infty}$ such that

$$\mathcal{V}_n \subset \mathcal{V}_{n+1}, \qquad \bigcap_{n \in \mathbb{Z}} \mathcal{V}_n = \{0\}, \qquad \text{and} \qquad \bigvee_{n \in \mathbb{Z}} \mathcal{V}_n = L^2\left(\mathbb{R}^d\right) \qquad (B.3)$$

is said to offer a *resolution*. (To stress the *variety* of spaces in this telescoping family, we often use the word *multiresolution*.) Here, the symbol $\bigvee$ denotes the closed linear span. In pictures, the configuration of subspaces looks like as illustrated in Figs. B.1 and B.2.

While the scale of resolution spaces in the figure refers to the case of subspaces in L^2, the same idea (*multiresolutions*) is also used for "resolution" of other kinds of data, e.g., the resolution of *digital* images; a case we shall discuss in detail in Chapters 3 and 6. In the present case, a base-point resolution subspace is chosen. It is indexed as level 0, and it is generated by a single function φ and its integral translates. The coarser resolution spaces are smaller, and the higher resolution spaces, successively bigger. The function φ is called the *scaling function* of the father function.

Remember when shopping for a new digital camera: just as important as the resolutions themselves (as given here by the scale of closed subspaces $\mathcal{V}_n$) are the associated spaces of *details*. As expected, the details of a signal represent the relative complements between the two resolutions, a *coarser* one and a more *refined* one. **Starting with the Hilbert-space approach**

to signals, we are led to the following closed subspaces (relative orthogonal complements):

$$\mathcal{W}_n := \mathcal{V}_{n+1} \ominus \mathcal{V}_n = \{f \in \mathcal{V}_{n+1} : \langle f \mid h \rangle = 0, \ h \in \mathcal{V}_n\}, \tag{B.4}$$

and the signals in these intermediate spaces $\mathcal{W}_n$ then constitute the amount of detail which must be added to the resolution $\mathcal{V}_n$ in order to arrive at the next refinement $\mathcal{V}_{n+1}$. In Figure B.2, the intermediate spaces $\mathcal{W}_n$ of (B.4) represent incremental details in the resolution [Jor03].

Here, we study *intermediate spaces*, coming from the scale of resolution subspaces in Figure B.1. These intermediate spaces correspond to *details*, (differences between pairs of resolution subspaces). And again, the detail information may refer to standard L^2 wavelet algorithms, or to more general data, e.g., arising from digital image algorithms. The latter may be grayscale or color, as explained in Chapter 3. Here, we first use dyadic L^2 wavelets, and wavelets with scale number N. We select a base-point resolution, and we identify wavelet generators ψ_i as generators for the *intermediate detail space*, called mother functions. With the use of translations, and iterated scaling with powers of N applied to wavelet generators ψ_i, we then obtain a wavelet basis for L^2 [Jor03].

The simplest instance of this is the one which Haar discovered in 1910 [Haa10] for $L^2(\mathbb{R})$. There, for each $n \in \mathbb{Z}$, $\mathcal{V}_n$ represents the space of all step functions with step size 2^{-n}, i.e., the functions f on $\mathbb{R}$, which are constant in each of the dyadic intervals $j2^{-n} \le x < (j+1)2^{-n}$, $j = 0, \dots, 2^n - 1$, and their integral translates, and which satisfy $\|f\|^2 = \int_{-\infty}^{\infty} |f(x)|^2 \, dx < \infty$. The inner product for Haar is the familiar one,

$$\langle f \mid h \rangle = \int_{-\infty}^{\infty} \overline{f(x)} \, h(x) \, dx, \tag{B.5}$$

and similarly for our present $L^2(\mathbb{R}^d)$, with the modification that the integration is now over $\mathbb{R}^d$.

An operator U in a Hilbert space is *unitary* if it preserves the norm, or equivalently, the inner product. Unitary operators are invertible and $U^{-1} = U^*$ where the $*$ refers to the adjoint. Similarly, the orthogonality property for a projection P in a Hilbert space may be stated purely algebraically as $P = P^2 = P^*$. The adjoint $*$ is also familiar from matrix theory, where $(A^*)_{i,j} = \overline{A_{j,i}}$: in other words, the $*$ refers to the operation of transposing and taking the complex conjugate. In the matrix case, the norm on $\mathbb{C}^n$ is $\left(\sum_k |x_k|^2\right)^{1/2}$.

For Haar's case, we can scale between the resolutions using $f(x) \mapsto f(x/2)$, which represents a dyadic scaling. To make it unitary, take

$$U = U_2 \colon f \longmapsto 2^{-\frac{1}{2}} f\left(\frac{x}{2}\right), \tag{B.6}$$

which maps each space $\mathcal{V}_n$ onto the next coarser subspace $\mathcal{V}_{n-1}$, and $\|Uf\| = \|f\|$, $f \in L^2(\mathbb{R})$. This can be stated geometrically, using the respective *orthogonal projections* P_n onto the resolution spaces $\mathcal{V}_n$, as the identity

$$U P_n U^{-1} = P_{n-1}. \tag{B.7}$$

And (B.7) is a basic geometric reflection of a self-similarity feature of the cascades of wavelet approximations. It is made intuitively clear in Haar's simple, but illuminating, example, included in the following. The important fact is that this geometric self-similarity, in the form of (B.7), holds completely generally. Moreover, it serves as a tool for generating new wavelets, and for analyzing them. A crucial observation Haar made in his 1910 paper was that the box wavelet of Figure B.3 is actually singly generated. Without making it explicit, Haar also further noticed a special case of what is now called *multiresolution analysis* (MRA). Haar considered the two functions

$$\varphi = \chi_{[0,1)} \quad \text{and} \quad \psi = \chi_{[0,1/2)} - \chi_{[1/2,1)} \tag{B.8}$$

shown in Figures B.3(a) and B.3(b), where we use χ to denote the indicator function. With these two functions φ and ψ, it is clear that

$$\mathcal{V}_0 = \bigvee \{\varphi(\cdot - k) : k \in \mathbb{Z}\} \quad \text{and} \quad \mathcal{W}_0 = \bigvee \{\psi(\cdot - k) : k \in \mathbb{Z}\},$$

where $\bigvee$ denotes the closed linear span of the functions inside $\{\ \}$. Similarly, using the translation operators

$$(T_y f)(x) := f(x - y), \quad f \in L^2(\mathbb{R}), \ y \in \mathbb{R},$$

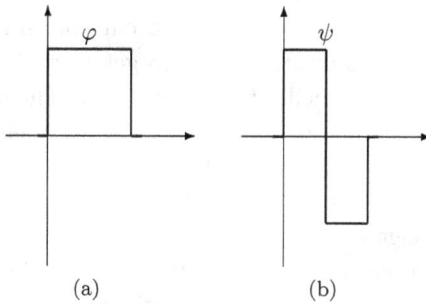

(a)   (b)

Fig. B.3   Haar wavelet functions.

and the relation

$$UT_yU^{-1} = T_{2y}, \qquad y \in \mathbb{R}, \tag{B.9}$$

we get

$$\begin{aligned}
\varphi(x) &= \varphi(2x) + \varphi(2x - 1), \\
\psi(x) &= \varphi(2x) - \varphi(2x - 1),
\end{aligned} \tag{B.10}$$

and, for each $n \in \mathbb{Z}$,

$$\mathcal{V}_n = \bigvee_{k \in \mathbb{Z}} \varphi(2^n x - k), \tag{B.11}$$

representing the closed subspace generated by all the $\mathbb{Z}$-translates as specified, and

$$\mathcal{W}_n = \bigvee_{k \in \mathbb{Z}} \psi(2^n x - k). \tag{B.12}$$

B.2   From Haar to Daubechies

While the identities (B.10), and the properties sketched in (B.11), are in fact clear from inspection of the shapes in Figure B.3, the first surprise in wavelet theory is that smooth wavelet shapes, represented by differentiable functions φ and ψ of compact support, are also possible, and with the functions satisfying the exact same resolution properties which were first noticed in a very special (non-smooth) case by Alfred Haar. We now turn to this crucial issue.

The issue of differentiability of wavelets is a rather large subject. We will only be able to touch on it here. Our viewpoint is that when the support size is specified, then we are able to display a corresponding variety of wavelets. The next step then is to identify the most differentiable specimens in the variety. This is in fact an area with current and exciting research, much of it dictated by applications; but to get started, we first need easy matrix formulations which facilitate computations. The tools are somewhat technical. Here is a sample of them, identified by their technical names: (i) the vanishing moment method (based on polynomial factorization); (ii) the joint spectral radius method (a clear-cut test, but difficult to apply); (iii) the dominant eigenvalue test (sketched in Figure C.3); (iv) the spin vector test. This last one amounts to writing the so-called polyphase matrix function $z \mapsto G(z)$ as a product of matrix functions $zp + p^{\perp}$ where p is a rank-one projection in some $\mathbb{C}^N$, i.e., p is the projection onto some

$\mathbf{v} \in \mathbb{C}^N$, $\|\mathbf{v}\| = 1$. By adjusting the configuration of vectors $\mathbf{v}$ contributing to the product factorization for $G(z)$, the more differentiable wavelets can be identified with a search algorithm. The graphics around Figures C.3 and C.4 may help the reader to visualize method (iv). Figure C.4 picks out a particular sample of configurations of two spin vectors. We explain in what follows how variations of a single unit vector $\mathbf{v}$ in $\mathbb{C}^2$ describe a variety of dyadic wavelets supported from 0 to 3 on the x-axis, while two independently moving spin vectors, i.e., unit vectors $\mathbf{v}_1$ and $\mathbf{v}_2$ in $\mathbb{C}^2$, describe the variety of wavelets supported in $[0, 5]$. Hermitian projections, especially finite-dimensional and one-dimensional ones, along with issues arising from their interactions and compositions, form a chapter of lore in complex geometry and in complex Hilbert space, as it is used in quantum theory. This framework describes the wavelet varieties perfectly: First recall that the complex n-dimensional subspaces in $\mathbb{C}^N$ are viewed as, and are by definition, points in the Grassmannian $G(n, N)$. And, starting with Wolfgang Pauli, $n = 1$, $N = 2$, it is popular to identify points in three pairwise isomorphic manifolds, (A)–(C) described as follows (for details, see, e.g., [BJ02b]): (A) $G(1, 2)$, (B) the two-sphere S^2, which goes under the name "the Bloch sphere" in physics circles (to Pauli, a point in S^2 represents the state of an electron, or of some spin-1/2 particle, and the points in the open ball inside S^2 represent mixed states), and finally, (C) equivalence classes of unit vectors in $\mathbb{C}^2$, where equivalence of vectors $\mathbf{u}$ and $\mathbf{v}$ is defined by $\mathbf{u} = c\mathbf{v}$ with $c \in \mathbb{C}$, $|c| = 1$. With this viewpoint, a one-dimensional projection p in $\mathbb{C}^N$ is identified with the equivalence class defined from a basis vector, say $\mathbf{u}$, for the one-dimensional subspace $p(\mathbb{C}^N)$ in $\mathbb{C}^N$. A nice feature of the identifications, $N = 2$, is that, if the unit vectors $\mathbf{u}$ are restricted to $\mathbb{R}^2$, sitting in $\mathbb{C}^2$ in the usual way, then the corresponding real submanifold in the Bloch sphere S^2 is the great circle: the points $(x, y, z) \in S^2$ given by $y = 0$. To Pauli, S^2, as it sits in $\mathbb{R}^3$, helps clarify the issue of quantum observables and states. Pauli works with three spinmatrices for the three coordinate directions, x, y, and z. They represent observables for a spin-1/2 particle. States are positive functionals on observables, so Pauli gets a point in $\mathbb{R}^3$ as the result of applying a particular state to the three matrices. The pure states give values in S^2. Recall that pure states in quantum theory correspond to rank-one projections, or to equivalence classes of unit vectors. When this viewpoint is applied to the wavelet formulation, we get an economical way of identifying the essential wavelet numbers, i.e., the masking coefficients for the subdivisions of wavelet theory. Most of the geometry we recall here carries over *mutatis mutandis* to unit vectors in $\mathbb{C}^N$. In the wavelet case, the number N represents a fixed scaling. Now the Grassmannian is $G(1, N)$, $N > 2$, but the analogue of the Bloch sphere is a little more complicated. It is worked out in geometry books, such as that of R.O. Wells, Jr. [Wel80]. What *is* considerably more complicated mathematics is the parameterization of the variety of states corresponding to several particles. It involves the notion of *entanglement* from quantum theory. For the present purpose, we use a finite set of projections p corresponding to the special $(zp + p^\perp)$-factors in a factorization of our polyphase matrix $G(z)$, describing a particular wavelet.

The existence of certain differentiable wavelets was discovered in the 1980s, see especially [Dau92]. I. Daubechies's, Y. Meyer's, and A. Cohen's pioneering discoveries opened a floodgate: Since then, new and powerful methods and techniques have emerged that provide constructive algorithms for optimal choices of resolutions and wavelets. These more recent methods, motivated by signal processing, include (i)–(iv) above, and they enable existence generally of signals of *compact support*, represented by compactly supported functions in $L^2(\mathbb{R}^d)$. They are also at the core of the many success stories of wavelet algorithms [Coh03b, Mey00b]. When the first wavelet constructions came out in the early 1980s, their significance was in fact not readily accepted or understood, in some cases not believed. Perhaps the first example of J.-O. Strömberg [Str92] was stillborn! What really opened up the subject, and enriched both theory and applications, was the connection to signal processing. From that came the multiresolutions, the pyramid algorithms, and the applications to data compression, to still image encoding, and more. Connections to signal processing are made in [Dau92] and especially in [Mal09]. Clearly the geometric relations (B.10), so transparent from Figure B.3(a) and B.3(b), invite the following generalization. They take the form of subdivision operations with masking coefficients, popular in numerical analysis, as follows:

$$\varphi(x) = \sqrt{|\det A|} \sum_{k \in \mathbb{Z}^d} a_k \varphi(Ax - k), \tag{B.13}$$

$$\psi_i(x) = \sqrt{|\det A|} \sum_{k \in \mathbb{Z}^d} b_k^{(i)} \varphi(Ax - k), \qquad i = 1, \ldots, |\det A| - 1. \tag{B.14}$$

The numbers (a_k) in (B.13) are called *masking coefficients* because of the use of (B.13) in graphics algorithms: there the a-numbers represent the masks in the successive subdivision steps of the algorithm. Here, (a_k) and $(b_k^{(i)})$ are scalar sequences indexed by the lattice $\mathbb{Z}^d$, and A is a fixed $d \times d$ matrix over $\mathbb{Z}$. We will assume that the eigenvalues λ of A satisfy $|\lambda| > 1$. This generalizes Haar's expansive scaling $x \to 2x$. It is known generally that the wavelets of *compact support* may be described this way using specific systems (a_k) and $(b_k^{(i)})$ of *finite* sequences, i.e., sequences which are identically zero outside a finite subset, Λ say, of $\mathbb{Z}^d$: specifically, $a_k = 0$ for all $k \in \mathbb{Z}^d \backslash \Lambda$, and similarly for the sequence $(b_k^{(i)})_{k \in \mathbb{Z}^d}$, $i = 1, \dots, |\det A| - 1$. Iteration algorithms are used in the solution to the system (B.13)–(B.14). When solutions φ and ψ_i to this system can be found in $L^2\left(\mathbb{R}^d\right)$, then the starting point of the hierarchical wavelet algorithm is the recursive buildup of the subspaces $\mathcal{V}_n$ and $\mathcal{W}_n$ of (B.3)–(B.4) with the use of formulas (B.16)–(B.18). This can be done with finite matrix algorithms known as subdivision algorithms; see [BJ02b], [Mal09], and [Wic94]. The recursive construction may be visualized in the next example, but in a context of fractals [Jor03].

B.3 Groups of Wavelets

The formulas (B.6) and (B.9) from Haar carry over to the general case as follows. Consider a $d \times d$ invertible matrix A over $\mathbb{Z}$. With scaling in the form

$$U : f \longmapsto |\det A|^{-1/2} f\left(A^{-1}x\right), \qquad x \in \mathbb{R}^d,\ f \in L^2\left(\mathbb{R}^d\right), \qquad \text{(B.15)}$$

we have

$$U T_y U^{-1} = T_{Ay}, \qquad y \in \mathbb{Z}^d. \qquad \text{(B.16)}$$

Note that $Ay \in \mathbb{Z}^d$ for $y \in \mathbb{Z}^d$ if the matrix A is integral. Moreover, generalizing (B.11)–(B.12), we get

$$\mathcal{V}_n = \bigvee \varphi\left(A^n x - k\right), \qquad \text{(B.17)}$$

$$\mathcal{W}_n = \bigvee_{\substack{k \in \mathbb{Z}^d \\ i = 1, \dots, |\det A| - 1}} \psi_i\left(A^n x - k\right), \qquad \text{(B.18)}$$

provided the coefficients in (B.13)–(B.14), called masking coefficients, satisfy the following orthogonality relations (where we have set $b_k^{(0)} := a_k$):

$$\sum_{k \in \mathbb{Z}^d} \left| b_k^{(i)} \right|^2 = 1 \qquad \text{for all } i \qquad \text{(normalization), and}$$

(B.19)

$$\sum_{k \in \mathbb{Z}^d} \overline{b_k^{(i)}} \, b_{k-Al}^{(j)} = 0 \qquad \text{for all } i \neq j, \text{ and } l \in \mathbb{Z}^d \qquad \text{(orthogonality).}$$

(B.20)

Introducing the d-torus as $\mathbb{T}^d = \mathbb{R}^d / \mathbb{Z}^d$, it is shown in [BJ02b] that the system (B.19)–(B.20) is equivalent to specifying a function from $\mathbb{T}^d$ into the $N \times N$ unitary matrices, where $N = |\det A|$. Since the matrix A is fixed and integral, the matrix multiplication on $\mathbb{R}^d$, $x \to Ax$, passes to the quotient $\mathbb{T}^d = \mathbb{R}^d / \mathbb{Z}^d$, and the induced mapping T_A is N-to-1. We now describe an isomorphism between the wavelet systems (B.19)–(B.20) and the *loop group elements*. (The group $\mathfrak{G}$ of measurable functions from $\mathbb{T}^d$ to the matrix group $U_N(\mathbb{C})$ is called the *loop group*.) Using a finite Fourier transform, we get a *function-valued inner product*, for functions on $\mathbb{T}^d$:

$$\langle p \mid q \rangle_A (z) = \frac{1}{N} \sum_{\substack{w \in \mathbb{T}^d \\ T_A w = z}} \overline{p(w)} \, q(w), \qquad z \in \mathbb{T}^d, \qquad \text{(B.21)}$$

where p and q vary over all scalar-valued functions on $\mathbb{T}^d$. In Haar's case, the sum on the right-hand side in (B.21) is over $\pm \sqrt{z}$. The application is to the case when these functions are

$$m^{(j)}(z) := \sum_{k \in \mathbb{Z}^d} b_k^{(j)} z^k, \qquad j = 1, \ldots, N-1,$$

where $z^k := z_1^{k_1} z_2^{k_2} \cdots z_d^{k_d}$, and $z \in \mathbb{T}^d$. Using the standard inner product on $\mathbb{C}^N$, it is therefore natural to extend (B.21) to the case of vector-valued functions $p \colon \mathbb{T}^d \to \mathbb{C}^N$, and revise (B.21) to

$$\langle p \mid q \rangle_A (z) = \frac{1}{N} \sum_{\substack{w \in \mathbb{T}^d \\ T_A w = z}} \langle p(w) \mid q(w) \rangle_{\mathbb{C}^N}. \qquad \text{(B.22)}$$

The key step in the identification of the solutions to (B.19)–(B.20) with the group $\mathfrak{G}$ of matrix functions $G \colon \mathbb{T}^d \to U_N(\mathbb{C})$ is now the following lemma.

Lemma B.2. *The group $\mathfrak{G}$ of all functions $G\colon \mathbb{T}^d \to \mathrm{U}_N\,(\mathbb{C})$ from the d-torus into all the $N \times N$ unitary matrices acts transitively on vector functions m on $\mathbb{T}^d$ as follows: $m \mapsto m^G$, where $m^G\,(z) := G\,(T_A z)\,m\,(z)$, and the pointwise product is matrix times column vector.*

Note (recalling that the inner product (B.22) takes values in functions on $\mathbb{T}^d$) that this is a *unitary* action relative to the functional inner product (B.22) in the sense that the unitarity identity $\left\langle m^G \mid p^G \right\rangle_A = \left\langle m \mid p \right\rangle_A$ holds for all $G \in \mathfrak{G}$. The term

$$\left\| m\,(z) \right\|^2 = \left\langle m\,(z) \mid m\,(z) \right\rangle_{\mathbb{C}^N} = \sum_{k=0}^{N-1} \left| m_k\,(z) \right|^2$$

measures the total contribution to a subband filter system, the N frequency bands being indexed by the cyclic group of order N, and the summation taken over the individual bands. Hence, the lemma states that this total contribution is constant under the specified action of the group $\mathfrak{G}$ of unitary matrix functions, the so-called polyphase matrices. For more details on the terminology of filters and matrix functions, see chapters 3 and 4.

If A is a given scaling matrix, i.e., a $d \times d$ matrix over $\mathbb{Z}$ with eigenvalues $|\lambda| > 1$, and $N := |\det A|$, then the corresponding filters m have N subbands, and are considered as functions from $\mathbb{T}^d$ to $\mathbb{C}^N$. With a slight abuse of notation, we say that m is a *quadrature mirror filter* (QMF) if $\left\langle m_i\,(z) \mid m_j\,(z) \right\rangle_A = \delta_{i,j}$, $z \in \mathbb{T}^d$, where $\left\langle \cdot \mid \cdot \right\rangle_A$ is the form in (B.21). It follows from the lemma, and an easy calculation, that if m and m' are any two given QMFs corresponding to the same scaling matrix A, then the functions

$$G_{i,j}\,(z) = \left\langle m_j\,(z) \mid m_i'\,(z) \right\rangle_A, \qquad z \in \mathbb{T}^d,$$

are matrix entries of a polyphase matrix G, i.e., $G \in \mathfrak{G}$, and $m' = m^G$. In other words, G transforms the first QMF m into the second m', and G is determined uniquely from m and m' [Jor03].

B.4 Examples

The significance of the approach via matrix functions is that it offers easy formulas for the numbers which encode our wavelet functions, φ, ψ (if the scale number N is 2), and φ, ψ_1, ..., ψ_{N-1} in general. Haar's two functions φ and ψ are supported in the unit interval $[0, 1]$, and Daubechies's father function φ and mother function ψ are supported in $[0,3]$. Then there is a next generation, still for $N = 2$, where φ and ψ are both supported in $[0,5]$. In terms of the matrix function $G(z)$, the case of $[0,1]$ is the constant function $G(z) = H = \dfrac{1}{\sqrt{2}} \begin{pmatrix} 1 & 1 \\ 1 & -1 \end{pmatrix}$, the next one is of the form $G(z) = H \left(zp + p^{\perp} \right)$ where p is a rank-1 projection in $\mathbb{C}^2$ and $p^{\perp} = I - p$. As p varies, we get the wavelet functions with support in $[0,3]$. With two rank-1 projections p_1, p_2 in $\mathbb{C}^2$ and $G(z) = H \left(zp_1 + p_1^{\perp} \right)\left(zp_2 + p_2^{\perp} \right)$, we get the wavelet functions supported in $[0,5]$. The simplest way of reading off the matrix function G which corresponds to a system of masking coefficients $\begin{pmatrix} a_0 & a_1 & a_2 & a_3 \\ b_0 & b_1 & b_2 & b_3 \end{pmatrix}$ is to set $A = \begin{pmatrix} a_0 & a_1 \\ b_0 & b_1 \end{pmatrix}$, $B = \begin{pmatrix} a_2 & a_3 \\ b_2 & b_3 \end{pmatrix}$, and $G(z) = A + zB$. The reader will be able to construct the higher-order cases by generalization, or he/she may consult [BJ02b].

B.4.1 *The world of the spectrum*

The theme of this section is that central properties of wavelets depend on the spectrum of a certain operator R, named after David Ruelle [Rue69], and also called the wavelet-transfer operator. See (C.9) for details. **Illustration:** Spread out over the area of Figure B.4, you have a family of wavelets, each generating function (scaling function) φ determined by two parameters. Having the algorithm for the scaling function, an extra step automatically gets us the wavelet itself. There is a coordinate system of wavelets on the interval $[0,5]$ with their associated scaling functions, see Figure B.4. Since vectors move invariantly under unitary transformations, it is reasonable to expect that varieties of wavelets may be parameterized in a way which is analogous to how we do the classical unitary groups. This will be illustrated in the following graphics experiment.

B.4.2 *Wavelets on the interval $[0,5]$ and their associated scaling functions*

The scaling functions φ for these wavelets, one for each sample point represented in Figure B.4, are pictured in this wall decoration, i.e., in Figures B.1, B.2 and B.3. The algorithm used in generating each of the functions φ in the little rectangles of Figures B.1 and B.2 will be outlined in the discussion Equation (B.23). However, Figure B.4 is in fact sampling

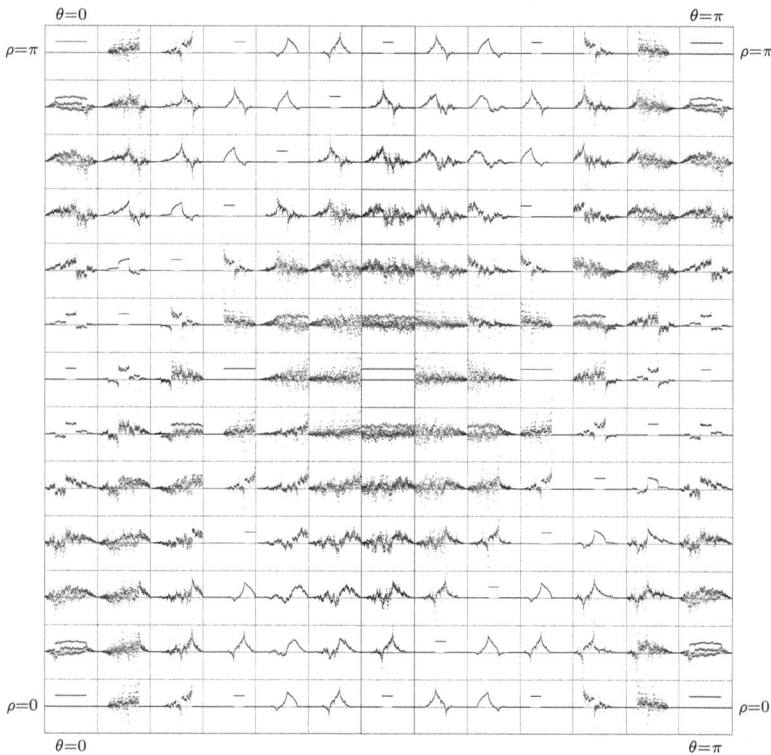

Fig. B.4 Two-parameter family of wavelet scaling functions. For detailed explanation of the generation of the family of scaling functions, see Section B.4.2.

from a two-parameter variety of orthogonal wavelets. (It is only a variety, because there are isolated points where the strict orthogonality property degenerates!) More precisely, we are looking at a genus-2 surface, but with deletion of these degenerates, or singular points: In the overall landscape of Figure B.4, these points are visibly represented by little flat scaling functions φ with support of length 3 or 5. (Remember the "honest" Haar wavelets are supported on an interval of unit length!) What is more significant, and also clearly visible in Figure B.4, is the fact that "most of" the functions φ in the variety are pretty bad!! The nicest that can be said about them is that they are L^2; and that is not all bad. (Remember the Devil's staircase function from Figure C.1 with its support on the Cantor set! As a scaling function, this "horrible" φ, it is not even locally integrable.)

So how *do* we find the nicer wavelets in a variety, those with more derivatives? There are two popular approaches: (i) Following [Dau92], look for vanishing moments. They can be found from factorization of certain polynomials, the Daubechies polynomials. Or: (ii) Do factorization of the polyphase matrix! For that, there is a handy operator R; it is a transfer operator, see (C.9). The spectrum of R is pictured in Figure C.3, the eigenvalues of R are functions of the two parameters that refer to Figure B.4. In the valleys of the mountainous landscape of level-surfaces from Figure C.3, i.e., the eigenvalues as they depend on two parameters, is where we pick out the nicest of the functions in the chaos of possibilities from the wavelets in Figure B.4. The operator R is nowadays called the wavelet-transfer operator, but it was studied first by David Ruelle in the 1960s in connection with a completely different problem (the phase transition question of quantum statistical mechanics [Rue69]), and later in chaos theory [Rue02].

It is convenient for each $k = 1, 2, \ldots$ to look at the family of all wavelet functions φ and ψ supported in the interval $[0, 2k+1]$ on the x-axis. Within the family of all orthogonal wavelets supported in $[0, 2k+1]$, we look for vanishing moments of order j where $1 \leq j \leq k$. We know how to find subvarieties $S_j \supset S_{j+1}$ where S_j consists of the functions ψ in the family such that ψ has vanishing moments of order j. As j increases, S_j contains wavelets of higher and higher orders of differentiability. Testing for smoothness in each S_j can be done with the so-called joint spectral radius (JSR) test. It is a JSR computed for two non-commuting 2×2 matrices built in a simple way from the system of masking coefficients $a_0, a_1, a_2, \ldots,$ a_{2k+1}. But since the 2×2 matrices are non-commuting, the JSR is hard to compute.

An alternative test for smoothness involves the Ruelle operator R. For $k = 2$, the interval is $[0, 5]$, and, for real-valued wavelets, the full variety is given by the (θ, ρ) parameters. The eigenvalues of $R_{\theta, \rho}$ may be ordered as follows:

$$1 \geq |\lambda_1(\theta, \rho)| \geq |\lambda_2(\theta, \rho)| \geq \cdots.$$

There is a simple known function s from $\mathbb{R}_+$ to $\mathbb{R}_+$ such that if $|\lambda_1(\theta, \rho)| < c$, then the two functions $\varphi_{\theta, \rho}$ and $\psi_{\theta, \rho}$ are in the Sobolev space $\mathcal{H}^{s(c)}$. Recall that the Sobolev exponent $s(c)$ is a measure of differentiability. But algorithms for calculating the best s are few, and not especially efficient. Nonetheless, continuity and differentiability of the wavelet functions are of critical importance in applications: The *a priori* estimates which give the best wavelet algorithm in JPEG 2000 are done in spaces of functions of bounded variation, and they depend on the smoothness of the wavelets that are used.

In Chapters 1–2 in [BJ02a] and in [Tre], it is pointed out that families of compactly supported wavelets admit a group-theoretic formulation. When this idea is specialized to the case of multiresolution wavelets which have both the scaling function (father function φ) and the wavelet generator (mother function ψ) itself supported in the fixed interval from 0 to 5, then the full variety of possibilities may be described by two independently varying unit vectors in $\mathbb{C}^2$, in coordinates, $\mathbf{v} = (v_1, v_2) \in \mathbb{C}^2$, $\|\mathbf{v}\|^2 = |v_1|^2 + |v_2|^2 = 1$. Unit vectors in $\mathbb{C}^2$ define pure quantum mechanical states. The latter may be parameterized by points on the (Bloch) sphere S^2, i.e., points $(x, y, z) \in \mathbb{R}^3$, $x^2 + y^2 + z^2 = 1$. If σ_x, σ_y, and σ_z are the Pauli spin matrices $\left(\begin{smallmatrix} 0 & 1 \\ 1 & 0 \end{smallmatrix}\right)$, $\left(\begin{smallmatrix} 0 & -i \\ i & 0 \end{smallmatrix}\right)$, and $\left(\begin{smallmatrix} 1 & 0 \\ 0 & -1 \end{smallmatrix}\right)$ in the three coordinate directions, then $(x, y, z) = (\langle \mathbf{v} \mid \sigma_x \mathbf{v} \rangle, \langle \mathbf{v} \mid \sigma_y \mathbf{v} \rangle, \langle \mathbf{v} \mid \sigma_z \mathbf{v} \rangle)$ is in S^2 if and only if $\|\mathbf{v}\| = 1$. For example, $\mathbf{v} = (\cos\theta, \sin\theta)$ in $\mathbb{C}^2$ corresponds to the point $(\sin 2\theta, 0, \cos 2\theta)$ on S^2. Hence, viewing S^2 as embedded in $\mathbb{R}^3$, the vector moves on a great circle on S^2 in the (x, z)-plane. Thus, in the example with two vectors of the form $(\cos\theta, \sin\theta)$, $(\cos\rho, \sin\rho)$ with the parameters θ and ρ varying independently, displayed in [BJ02a, (1.2.3)–(1.2.4)], the dependence of the masking coefficients $a_k(\theta, \rho)$ on 2θ and 2ρ is as expected: the wavelet parameter is not the vector $\mathbf{v}$ itself, but rather the subspace generated by $\mathbf{v}$, or the projection onto $\mathbf{v}$, so we get the same wavelet masking coefficients for $\mathbf{v}$ and $-\mathbf{v}$. The coefficients a_k are real valued for a similar reason. The two functions φ and ψ depend on variations in (θ, ρ) periodically, with period π in both parameters, so the whole variety may be represented in a square $[0, \pi) \times [0, \pi)$; and this is illustrated graphically in a supplement to [BJ02a], http://www.math.uiowa.edu/~jorgen/wavelet_motions.pdf. It may be used as a flip-book of wavelets, or it may be explored with a mouse search on the screen.

While the functions φ and ψ are square integrable, they are continuous on $[0, 5]$ only for (θ, ρ) in a periodic subset of $\mathbb{R} \times \mathbb{R}$. (This subset is often identified by vanishing moments or by spectral conditions such as the JSR test; see [CGV99].) If we ask whether φ and ψ, as points in a function space, show continuous dependence on (θ, ρ), then that function-space continuity must be measured in mean square, i.e., in the metric defined by the L^2-norm of functions on $[0, 5]$. In this metric, continuity follows from [BJ02a, Theorem 2.5.8]; see details in what follows. Actually, there are a finite number of exceptions to the L^2-continuous dependence on (θ, ρ). They occur when the wavelet is one of the degenerate Haar cases. These are the Haar wavelets which do not define strict orthonormal bases, but only tight frames. They have φ of the form $\varphi = c\chi_I$ where I is a subinterval of $[0, 5]$ of length 3 or 5, and where $c = \frac{1}{3}$ in the first case, and $c = \frac{1}{5}$ in the second. Find them on Figure B.4! If the reader follows the moving wavelet pictures (on a printout or on the screen), we hope he/she will get some intuitive ideas of fundamental wavelet relationships. The rigorous mathematics relating the various continuity properties to cascade approximation, to moments, and to spectral estimates, is covered in much more detail in the following references, especially [BJ02a] and [CGV99].

Once the masking coefficients $a_0, a_1, \ldots, a_5$ are specified as functions of the parameters θ and ρ, the computation of cascade approximants of the scaling function φ is done with a series of Mathematica operations. The algorithm is designed to start with a Haar function, and the limit of the iteration will then be a scaling function $\varphi = \varphi^{(\theta, \rho)}$ depending on the two rotation angles θ, ρ, and satisfying

$$\varphi^{(\theta, \rho)}(x) = \sqrt{2} \sum_{k=0}^{5} a_k^{(\theta, \rho)} \varphi^{(\theta, \rho)}(2x - k). \tag{B.23}$$

The reptile features of the algorithm have the effect of producing fast cascading approximations to the limit function $\varphi^{(\theta, \rho)}$. The algorithm of the solution $\varphi^{(\theta, \rho)}$ to the scaling identity (B.23) then proceeds as follows (see [Jor01a], [Tre], Section 1.2 of [BJ02a] for details). The relation (B.23) is interpreted as giving the values of the left-hand $\varphi^{(\theta, \rho)}$ by an operation performed on those of the $\varphi^{(\theta, \rho)}$ on the right. A binary digit inversion transforms this into

$$\mathbf{f}_{k+1}'(x) = \mathbf{A}\mathbf{f}_k(x),$$

where $\mathbf{A}$ is the 2×3 matrix $\mathbf{A}_{p,q} = \sqrt{2} a_{5+p-2q}^{(\theta, \rho)} = \sqrt{2} \begin{pmatrix} a_4^{(\theta, \rho)} & a_2^{(\theta, \rho)} & a_0^{(\theta, \rho)} \\ a_5^{(\theta, \rho)} & a_3^{(\theta, \rho)} & a_1^{(\theta, \rho)} \end{pmatrix}$

constructed from the coefficients in (B.23), and $\mathbf{f}_j$ and $\mathbf{f}_j'$ are the vector functions

$$\mathbf{f}_j(x) = \begin{pmatrix} \varphi^{(\theta, \rho)}\left(x - \frac{2}{2^j}\right) \\ \varphi^{(\theta, \rho)}\left(x - \frac{1}{2^j}\right) \\ \varphi^{(\theta, \rho)}(x) \end{pmatrix}, \qquad \mathbf{f}_j'(x) = \begin{pmatrix} \varphi^{(\theta, \rho)}(x) \\ \varphi^{(\theta, \rho)}\left(x + \frac{1}{2^j}\right) \end{pmatrix}.$$

Iterations of this operation give values of an approximation to $\varphi^{(\theta,\rho)}$ on successively finer dyadic grids in the x variable. For an implementation of this computation in Mathematica, we let loctwont stand for the transpose of the coefficient matrix $\mathbf{A}$,[1] and normalize the Mathematica variables a0, a1, etc., to include the factor of $\sqrt{2}$ that appears in the cascade iteration:

```
loctwont[θ_, ρ_] :=
 N[Transpose[{{a4[θ, ρ], a2[θ, ρ], a0[θ, ρ]}, {a5[θ, ρ], a3[θ, ρ], a1[θ, ρ]}}]]

cascadestep[phitable_, θ_, ρ_] :=
 Flatten[Partition[Flatten[{0, 0, phitable, 0, 0}], 3, 1].loctwont[θ, ρ]]

wavelet[phistart_, itercount_, θ_, ρ_] :=
 Transpose[{Table[i (2^(-itercount)), {i, 0, 5 (2^itercount) - 5}],
   Nest[cascadestep[#, θ, ρ] &, phistart, itercount]}]
```

Each cascade step works with the list of values from the previous step (phitable), pads it with zeroes on left and right, works it up into a matrix by overlapping divisions (the Partition operation), does a matrix multiplication with a matrix of masking coefficients (loctwont), and reduces (with Flatten) the resulting matrix to a list again. At each stage, the implicit grid spacing is halved, and at the specified final iteration, the values of these grid points are associated with the elements of the list (using Table and Transpose), so that a plot of the scaling-function approximant can be made with other Mathematica operations such as ListPlot.

A direct implementation in Mathematica of the computation of the wavelet ψ from the scaling function φ by the formula [BJ02a, (2.5.25)] (with the functions and coefficients depending on the rotation angle parameters θ, ρ),

$$\psi^{(\theta,\rho)}(x) = \sqrt{2} \sum_{k=0}^{5} (-1)^k a_{5-k}^{(\theta,\rho)} \varphi^{(\theta,\rho)}(2x - k),$$

is (again with $\sqrt{2}$ subsumed in a0, a1, ...)

---

[1] So named by contraction of "local two-by-n transpose."

```
embedwavelet[phistart_, itercount_, θ_, ρ_] :=
 Flatten[{Table[0, {i, -(5 (2^itercount) - 5), -1}], Nest[cascadestep[#, θ, ρ] &, phistart,
    itercount], Table[0, {i, (5 (2^itercount) - 5) + 1, 2 (5 (2^itercount) - 5)}]}, 1]

avec[θ_, ρ_] := {a0[θ, ρ], a1[θ, ρ], a2[θ, ρ], a3[θ, ρ], a4[θ, ρ], a5[θ, ρ]}

waveletpsi[phistart_, itercount_, θ_, ρ_] :=
 Transpose[{Table[i (2^(-itercount-1)), {i, 0, 2 (5 (2^itercount) - 5)}],
    Table[Sum[((-1)^(5-k) avec[θ, ρ][[k + 1]] embedwavelet[phistart, itercount, θ,
        ρ][[j + k (2^itercount-1)]]], {k, 0, 5}], {j, 1, 2 (5 (2^itercount) - 5) + 1}]}]
```

But using the `Table` and `Sum` operations in this way is inefficient for
two reasons. First, the sum has, after suitable rearrangement, the form of a
sequence correlation that can be implemented with a fast Fourier transform.
And second, the way the operations above are interpreted by Mathematica,
the same scaling function (`embedwavelet`) is computed repeatedly for each
term of the sum, rather than being computed once and saved. Both of these
inefficiencies can be remedied by the use of the `ListCorrelate` operation,
which Mathematica implements internally by Fourier-transform methods.

```
correlatewavelet[phistart_, itercount_, θ_, ρ_] :=
 Flatten[Transpose[Partition[PadLeft[PadRight[Nest[cascadestep[#, θ, ρ] &, phistart,
    itercount], 6 (2^itercount) - 6], 11 (2^itercount) - 11], (2^itercount) - 1]]]

signavec[θ_, ρ_] := {-a0[θ, ρ], a1[θ, ρ], -a2[θ, ρ], a3[θ, ρ], -a4[θ, ρ], a5[θ, ρ]}

correlatewaveletpsi[phistart_, itercount_, θ_, ρ_] :=
 Transpose[{Table[i (2^(-itercount-1)), {i, 0, 11 (2^itercount) - 12}],
    Flatten[Transpose[Partition[ListCorrelate[signavec[θ, ρ],
        correlatewavelet[phistart, itercount, θ, ρ], {1, 1}, 0], 11]]]}]
```

The necessary rearrangement amounts to grouping the list of values into
a matrix and transposing the matrix. This is done, at both ends of the
computation, with the `Partition` and `Transpose` operations, followed by
`Flatten` to return to an ungrouped list. Note also the use of `PadLeft`
and `PadRight` to add zeroes to keep the different rows of the matrix
from getting mixed in the `ListCorrelate` operation; these could have
been used in the direct implementation as well, where zeropadding was
needed at the ends of the sum. The Fourier-transform method used
in Mathematica's implementation of the `ListCorrelate` operation gives
results numerically identical to those obtained by the direct calculation
with the `Table` and `Sum` operations, thanks in part to Mathematica's
implementation of `ListCorrelate` that uses a real transform method on
real data. For the wavelet functions computed for the flip book, the method
using `ListCorrelate` works about 500 times faster than the method using
`Table` and `Sum` [Jor03].

Appendix C

Georg Cantor's Chaos

Note that the generalization from the so-called *refinement equation* in the form (B.10) of Haar to the general case (B.13) is not just a jump in dimension, from 1 to d — that is not even the main point. The main issue is identifying the L^2 solutions to (B.13); then (B.14) comes along for free. To understand this, consider the seemingly trivial modification of (B.10) into

$$\varphi(x) = \frac{3}{2}\left(\varphi(3x) + \varphi(3x - 2)\right). \tag{C.1}$$

Forgetting *functions* as solutions φ, try *tempered distributions*. Then a Fourier transform of (C.1) yields

$$\hat{\varphi}(\xi) = e^{-i\xi/2} \prod_{n=1}^{\infty} \cos\left(\frac{\xi}{3^n}\right). \tag{C.2}$$

It can be checked that, up to normalization, this is the unique solution, and that φ is the singular Cantor measure, and in particular a tempered distribution, corresponding to the middle-third construction. Hence, this φ is the unique measure satisfying

$$\int_{\mathbb{R}} f(x)\, d\varphi(x) = \frac{1}{2}\left(\int_{\mathbb{R}} f\left(\frac{x}{3}\right) d\varphi(x) + \int_{\mathbb{R}} f\left(\frac{x+2}{3}\right) d\varphi(x)\right) \tag{C.3}$$

and φ is supported in the set of Figure C.1. The use of the Fourier transform in (C.2) makes perfectly good sense, even if φ is not a locally integrable function on $\mathbb{R}$. It would make sense even if we only knew *a priori* that φ were a compactly supported distribution: then for each $\xi \in \mathbb{R}$, we would define $\hat{\varphi}(\xi)$ as the distribution φ applied to the C^∞ function $x \mapsto e^{-i\xi x}$.

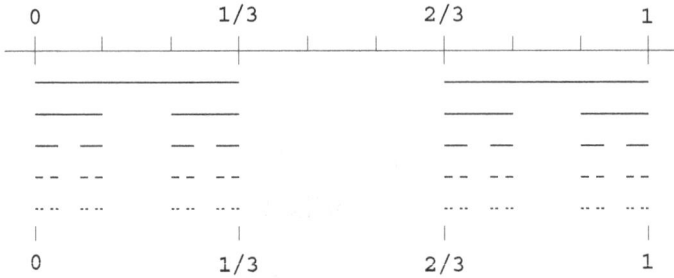

Fig. C.1 Iteration leading to Cantor-set support of φ in (C.1).

In actual fact, the iteration algorithm based on (C.3) which is illustrated in Figure C.1 shows that φ is the Cantor measure. Hence,

$$\hat{\varphi}\left(\xi\right) = \int e^{-i\xi x}\, d\varphi\left(x\right).$$

Example C.1. Cloud Nine as a reptile. (The name *reptile* refers to a tiling property which is self-reproducing in the sense that the picture repeats itself at all scales.) As an example of (B.13), take $d = 2$ and $A = \begin{pmatrix} 1 & 2 \\ -2 & 1 \end{pmatrix}$. Then $N = \det A = 5$, and the five lattice points $D = \left\{ \begin{pmatrix} 0 \\ 0 \end{pmatrix}, \begin{pmatrix} \pm 3 \\ 0 \end{pmatrix}, \begin{pmatrix} 0 \\ \pm 2 \end{pmatrix} \right\}$ represent all the five residue classes for the quotient group $\mathbb{Z}^2 / A\mathbb{Z}^2$. The formula $\varphi\left(x\right) = \sum_{d \in D} \varphi\left(Ax - d\right)$ is then a special case of the scaling Equation (B.13), and it is at the same time a natural extension of Haar's Equation (B.10). It has a solution $\varphi = \chi_{\mathbf{T}}$, and the compact set $\mathbf{T} \subset \mathbb{R}^2$ is the unique solution to the so-called reptile Equation $A\mathbf{T} = \bigcup_{d \in D} \left(\mathbf{T} + d\right)$; see also Figure C.2. It is shown in [BJ99] that $\mathbf{T}$ tiles $\mathbb{R}^2$ with translations chosen from the lattice $\Lambda = \left\{ \begin{pmatrix} l \\ 2m \end{pmatrix} : l, m \in \mathbb{Z} \right\}$.

---

[1]The name here is from [BJ99]; the picture represents a dynamical systems orbit of order 9.

[2]Take a geometric figure, replicate it five times, in the original location, three units to the right, two units above, two units below, and three units to the left, then shrink the whole collection to the same size as the original was, and rotate it by a "knight's move" angle. Repeat the process any number of times (here seven). If the initial figure is suitably chosen, the figures never overlap, and the collection tiles the plane by the same lattice at each stage of the iteration. Continuously varying colors assigned to the pieces in the order they were created show how the parts interpenetrate.

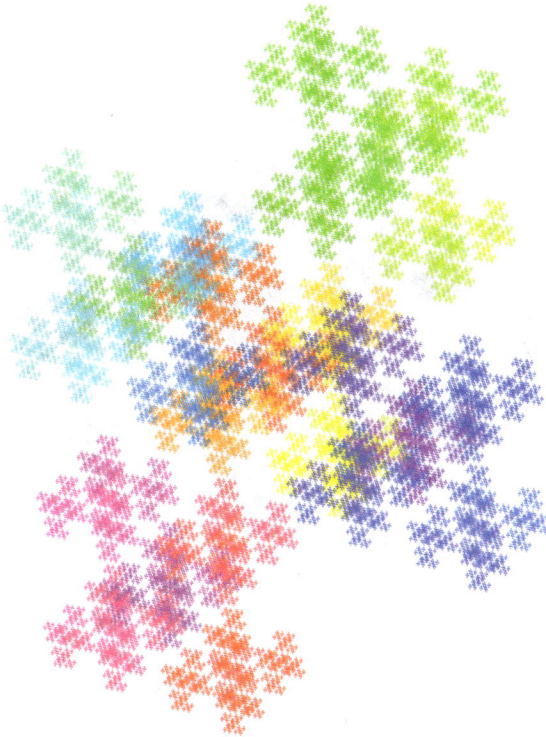

Fig. C.2 A Rep-Tile. The tile **T** from the "Cloud Nine" example[2]. The picture is a result of iterating five linear maps, all having an expansive scaling matrix of determinant 5. The result is an exotic "tiling" of the plane[3]. The boundary of the tile is a fractal, like the self-similar structures that occur in biological systems, in correlations of fluids at critical phase transitions, and in conditions of turbulent flow.

It follows from this that the measure of **T** is 2. Then **T** is a reptile or a generalized Haar wavelet. In general, the measure of a reptile is an integer. It is known, however, that in $\mathbb{R}^4$ there are 4-by-4 expansion matrices A over $\mathbb{Z}$ for which generalized Haar wavelets do not exist. In 4-D for special matrices A, it may be that the reptiles generated by this scheme might not tile by a lattice which would turn them into Haar wavelets.

We present a guide diagram in Figure C.4 corresponding to the (θ, ρ) plane of Figures C.2 and C.3, with dots for the point (and its periodic replicas) whose wavelet function and scaling function are displayed on the current page and line. The square outlined in black is the fundamental region $[0, \pi) \times [0, \pi)$, shown with labels in Figure C.4. There is a path,

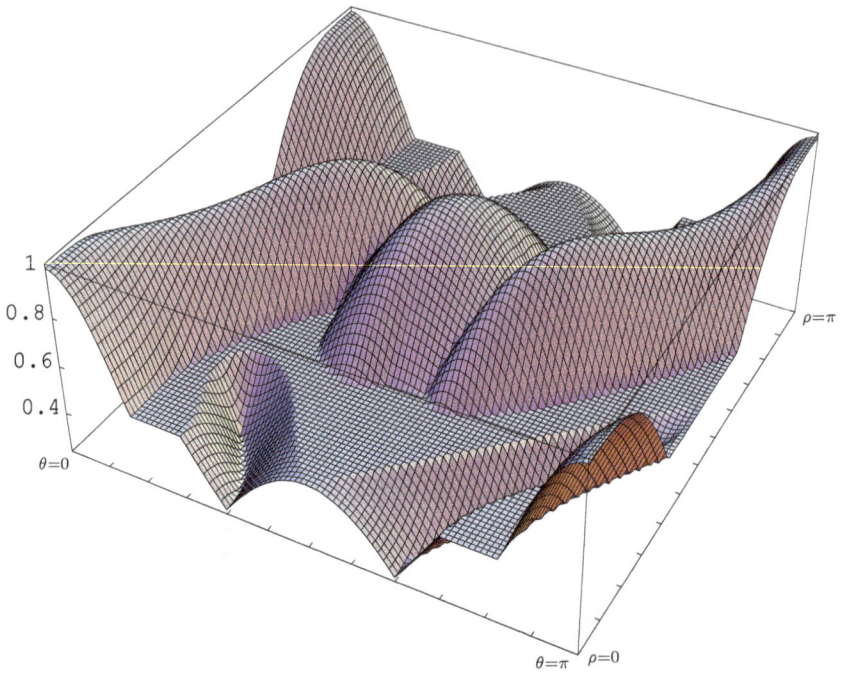

Fig. C.3   Eigenvalue of two-parameters family of Ruelle operators.

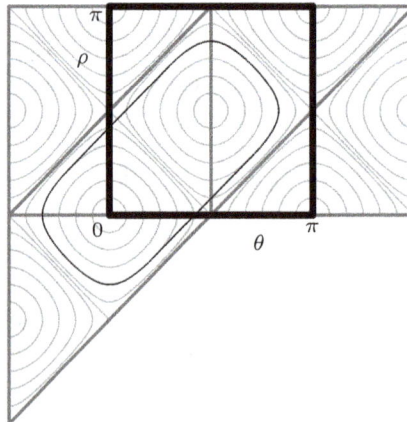

Fig. C.4   Guide diagram

part of which represents a vanishing-moment curve where the wavelet function $\psi(x)$ satisfies the moment condition $\int x\,\psi(x)\,dx = 0$, and the masking coefficients $a_0, \ldots, a_5$ satisfy the equivalent divisibility condition $m_0(z) = 2^{-\frac{1}{2}} \sum_{k=0}^{5} a_k z^k = (z+1)^2 p(z)$ for some polynomial p. The viewer familiar with [Jor01a] may find amusement in watching for the appearance in the sequence of the familiar Daubechies wavelets, as well as the single point in the sequence where the additional moment and divisibility conditions $\int x^2\,\psi(x)\,dx = 0$ and $m_0(z) = (z+1)^3 q(z)$ are met. In Figure B.2, you see variations of these wavelets, as the parameters move around and cover a period-square.

The part of the spectrum of the Ruelle operator R in (C.9) which determines the shape of each wavelet is a finite set of points. As each point in this part of the spectrum is a function of the two parameters, the eigenvalues may be ordered and presented in sketch as functions of two variables. The graphs represent surfaces starting with a top eigenvalue (a Perron–Frobenius eigenvalue equal to 1 for all values of the parameters). One of the surfaces, two steps down from the top, is pictured in Figure C.3. There is a constant eigenvalue $1/2$ that gives the appearance of mountains rising from lakes, or Christopher Columbus's hat, or a body in bathwater if you prefer! And then the top eigenvalue equal to 1 is the ceiling. Other eigenvalue surfaces in this series are shown in figures within the book [BJ02a], and one surface with branch cuts, used as a sort of frontispiece, looks like the "Half Dome" in Yosemite Valley. This closed-form eigenvalue, when compared to the eigenvalue surfaces sorted by absolute value, coincides with the second eigenvalue from the top in part of the parameter range, and with smaller eigenvalues in other parts of the range. The parameters representing the lower parts of the surfaces are important: e.g., they give algorithms that are faster. And wavelets that are more regular tend to be located in the region of the period-square which has a concentration of low-lying spectrum. Hence, the flat portions of the "bathwater" surface signify an abundance of "good" wavelets [Jor03].

C.0.1 *Haar Meets Cantor*

The scaling identity (B.13) depends on a choice of masking coefficients $\{a_k\}_{k \in \mathbb{Z}^d}$. This suggests a strategy for finding the scaling function (also called the father function) φ as a solution to an operator identity: the idea is to try to get φ as the limit of an iteration of the following operator.

$$S: f \longmapsto \sqrt{N} \sum_{k \in \mathbb{Z}^d} a_k f \left(Ax - k \right); \qquad \text{(C.4)}$$

i.e., $\varphi = \lim_{n \to \infty} S^n f$ with a suitable choice of starting point f for the iteration. As suggested by (B.13), an integration shows that a necessary condition for this to work is that

$$\sum_{k \in \mathbb{Z}^d} a_k = \sqrt{N}. \qquad \text{(C.5)}$$

This normalization is indeed satisfied for the two examples (B.10) and (C.1). In the first one (B.10), $d = 1$, $A = N = 2$, and the solution φ is the box function φ of Haar which is certainly in $L^2(\mathbb{R})$. In the second example (C.1) of Cantor, $d = 1$, $A = N = 3$, and the solution φ is now the Cantor measure — this measure, while a solution, is a singular measure, and, in fact, it is not even represented locally by an integrable function.

To understand the dichotomy between the case of Haar and that of Cantor, introduce the generating function m_0 for the masking coefficients in (B.13), i.e., $m_0(z) = \sum_{k \in \mathbb{Z}^d} a_k z^k$. For simplicity, specialize to $d = 1$. It will be convenient for us to consider m_0 as a function on $\mathbb{T}$ by restriction, viewing the torus $\mathbb{T} = \mathbb{R}/\mathbb{Z}$ as a circle in $\mathbb{C}$, i.e., as $\mathbb{T} = \{z \in \mathbb{C} : |z| = 1\}$. Then the normalization condition (C.5) reads

$$m_0(1) = \sqrt{N}. \qquad \text{(C.6)}$$

The other conditions (B.19)–(B.20) on the masking coefficients, i.e.,

$$\sum_{k \in \mathbb{Z}} \bar{a}_k a_{k - lN} = \delta_{0,l}, \qquad l \in \mathbb{Z}, \qquad \text{(C.7)}$$

then take the functional form

$$\frac{1}{N} \sum_{\substack{w \in \mathbb{T} \\ w^N = z}} \left| m_0(w) \right|^2 = 1. \qquad \text{(C.8)}$$

But this is an orthogonality condition, stated in either one of its two equivalent forms, and it is necessary for ensuring orthogonality in the ultimate construction of a wavelet basis for $L^2(\mathbb{R})$. In Haar's example, $N = 2$ and $m_0(z) = \frac{1+z}{\sqrt{2}}$. Setting $z = e^{-i\xi}$, it follows that (C.8) is equivalent to the familiar trigonometric identity $\cos^2 \xi + \sin^2 \xi = 1$. But, for Cantor's example, $N = 3$ and $m_0(z) = \frac{\sqrt{3}}{2}(1 + z^2)$, and now $\frac{1}{3} \sum_{w^3 = z} \left| m_0(w) \right|^2 = 3/2$.

Stated differently, if an operator R is defined as acting on functions h on $\mathbb{T}$ by

$$(Rh)(z) = \frac{1}{N} \sum_{\substack{w \in \mathbb{T} \\ w^N = z}} |m_0(w)|^2 h(w), \tag{C.9}$$

then the formula (C.8) takes the form $R\mathbf{1} = \mathbf{1}$, where $\mathbf{1}$ denotes the function on $\mathbb{T}$ which is constant and equal to 1. In the analysis of [Dau92] and [BJ02a], the operator R from (C.9) is called the wavelet transfer operator, or the Ruelle operator, and $\mathbf{1}$ is called the Perron–Frobenius eigenfunction. Hence, the significant distinction between the case of Haar and that of Cantor is the numerical size of the Perron–Frobenius eigenvalue: It is 1 in the first case and $3/2$ in the second. The operator R in (C.9) is called the *wavelet transfer operator* because of the analogy to the probabilistic view of the Perron–Frobenius matrix setting, and because of the probability content of condition (C.8): Looking at the right-hand side in (C.9), for each point $z \in \mathbb{T}$, there are N distinct solutions w to $w^N = z$, and each of the corresponding numbers $\frac{1}{N}|m_0(w)|^2$ signifies conditional probabilities for transfer from some w to z, where w is one of the N possibilities. This random-walk model helps us understand and compute the spectral theory of R, as pointed out in [BJ02a].

The Ruelle transfer operator R "counts" the accumulated contribution to the cascade approximation, i.e., to the scaling function φ, when the limit exists. But unless the Perron–Frobenius eigenvalue $\lambda_{\mathrm{PF}} = 1$, the limit "escapes" out from $L^2(\mathbb{R})$. In the middle-third-Cantor case, $\lambda_{\mathrm{PF}} = 3/2$, which accounts for φ in this case being a singular measure, not represented by a locally integrable function on $\mathbb{R}$. The Hausdorff dimension d in Cantor's case can be shown to be $d = \frac{\log 2}{\log 3}$.

Appendix D

Markov Chains and Generalized Wavelet Multiresolutions

In this appendix, we give some application examples of wavelet multiresolutions to Markov chains that are applied on fractals, specifically Julia set inspired by our results [JS18]. We show wavelet-like fractal example with multiresolutions.

D.1 Introduction and Setting

While there are already a number of approaches to harmonic analysis of fractals, "non-smooth" settings, we propose below a focus on a certain family of positive operators. They will serve as transfer operators. The novice might find the following references helpful: [Bal00, DJ06a].

In this appendix, we first develop the needed results on transfer operators, and the second part will be concrete applications. There are many justifications for the need of a constructive harmonic analysis of fractals; one is the discovery of Jorgensen–Pedersen that certain fractal L^2 spaces admit Fourier bases; while others do not. However, the lack of available Fourier bases in many examples suggests a need for alternative approaches.

The Cantor fractals are special cases of more general IFS systems. This chapter will deal with this more general framework. In addition to fractal Fourier analyses (fractals in the large), we shall also present multiresolution and wavelet techniques. In the work of Dutkay–Jorgensen, it was shown that the general affine IFS systems, even if not amenable to Fourier analysis, in fact do admit wavelet bases, and so in particular can be analyzed with the use of multiresolutions; reflecting the inherent self-similarity to the fractal under consideration. But this approach in fact depends on the use of certain transfer operators. The latter in turn ties in with intriguing new work on cascade algorithms, with an analysis of representations of non-commutative generators and relations (especially the Cuntz relations), as well as with certain stochastic processes; and we shall make connections to recent research on Markov processes, and to reproducing kernel theory [JS18].

D.2 Julia Sets

Definition D.1.

$$\sigma(z) = \frac{F(z)}{G(z)}, \tag{D.1}$$

where $\deg(\sigma)$ = maximum of degree of $F(z), G(z)$. Let $\tau_j : X_\sigma \to X_\sigma$ be defined to be

$$\sigma(\tau_j(z)) = z$$

$$\sigma_c(z) = z^2 + c.$$

With $\pm\sqrt{z-c}$, $(\pm\sqrt{z-c})^2 + c = z - c + c = z$.

Let $X = J_c$ be a Julia set and let $\sigma : X \to X$ where $X \subset \mathbb{C}$. $\{\tau_j\}_{j=1}^N$, $N := \deg \sigma$, and $\sigma \circ \tau_j = id_X$.

Example D.2. Fix $c \in \mathbb{C}$, $\sigma = \sigma_c$

$$\sigma_c(z) = z^2 + c$$

$$\{\tau_j\}_{j=1}^N : \pm\sqrt{z-c}$$

$$X_j := \tau_j(X), \quad X = \cup_j X_j \quad X_j \cup X_k = \phi_{j \neq k}.$$

which are partitions.

$$\sigma \circ \tau_j = id_X \quad \forall j = 1, \ldots, N.$$

Lemma D.3.

$$[m_j \circ \tau_k]_{j,k=1}^N = \begin{bmatrix} 1 & 0 & 0 & \cdots & 0 \\ 0 & 1 & 0 & \cdots & 0 \\ 0 & 0 & 1 & \cdots & 0 \\ \vdots & \vdots & & \ddots & \\ 0 & 0 & 0 & \cdots & 1 \end{bmatrix} = \delta_{jk}.$$

See [Jor01b, CG93].

Let

$$m = \begin{bmatrix} m_1 \\ m_2 \\ \vdots \\ m_N \end{bmatrix},$$

$$A = X \to U_N(\mathbb{C})$$

which is in $G(X, N)$.

$$A^* A = I_N$$

$$A^*(x) A(x) = I_N$$

which are unitary.

Then we have filter

$$m_j(z) := \chi_j(z) = \chi_{\tau_j(X)}(z)$$

$$A \in G(X, N), \quad A : X \to U_N(\mathbb{C}).$$

$$m \in \mathcal{MF} \quad \text{set } m^A(x) = A(\sigma(x)) m(x), \quad A = (A_{ij}), \quad m = (m_i) = \begin{bmatrix} m_1 \\ m_2 \\ \vdots \\ m_N \end{bmatrix}$$

D.3 Multiresolutions and Generalized Wavelet Representations

As is illustrated in [Jor06a], and the references given there; as well as in the papers [Dut04, DJ05, DJ06b, DPS14], there is a host of problems from analysis of fractals and more generally in stochastic analysis which lend themselves to the present multiresolution approach. Below, we discuss related wavelet representations.

Lemma D.4. *Let $(\Omega, \mathcal{F}, \mathbb{P})$ be a probablistic space, and let $\mathcal{A} : \Omega \to X$ be a random variable with values in a fixed measure space $(X, \mathcal{B}_X)$, then $V_{\mathcal{A}} f :=$ $f \circ \mathcal{A}$ defines an isometry. $L^2(X, \mu_A) \to L^2(\Omega, \mathbb{P})$ where μ_A is the law of A, i.e., $\mu_A(\Delta) := \mathbb{P}(\mathcal{A}^{-1}(\Delta))$, for all $\Delta \in \mathcal{B}_X$; and $V_{\mathcal{A}}^*(x) = \mathbb{E}_{\mathcal{A}=x}(\psi | \mathcal{F}_A)$ for all $\psi \in L^2(\Omega, \mathbb{P})$, and all $x \in X$.*

If $(\Omega, \mathcal{F}, \mathbb{P})$ is a solenoid probability span on $\Omega_X = \prod_{n=0}^{\infty} X$, we shall apply the lemma to the each vertex.

$\pi_n : \Omega_X \to X$ given by $\pi_n(x_0 x_1 x_2 \cdots) = x_n$, for all $n \in \mathbb{N}_0$, and the isometry span to π_n will simply be denoted V_n. The sigma-algebra given by π_n will be denoted $\mathcal{F}_n$. Let $(X, \mathbb{B})$ be fixed, let $Rf(x) = \int f(y)\mu(dy|x)$, $f \in \mathcal{F}(X, \mathcal{B})$ and $Rh = h$, i.e., $\mu(\cdots|x)$ is a probability space and $(X, \mathcal{B})$ a.e. $x \in X$. Suppose there exists an X and w such that $\int \mu(B|x)d\lambda(x) = \int_B wd\lambda$, for all $B \in \mathcal{B}_X$ then there exists a probability space $(\Omega, \mathbb{P})$ which is the all paths on $(X, \mathcal{B})$ such that

$$\int_\Omega (f_0 \circ \pi_0)(f_1 \circ \pi_1) \cdots (f_n \circ \pi_n)d\mathbb{P}$$

$$= \int_X f_0(x)R(f, R(f_2 \cdots R(f_n) \cdots)(x)d\lambda(x)$$

and $\mathbb{P} \circ \pi_1^{-1} = ((W \circ \pi_0)d\mathbb{P}) \circ \pi_0^{-1}$. Moreover,

$$\text{suppt}(\mathbb{P}) = \text{Sol}_\sigma(X) \iff$$
$$R[(f \circ \sigma)g] = fR(g), \quad \forall f, g \in \mathcal{F}(X, \mathcal{B}).$$

Suppose (R, λ) has the representation $(Rf)(x) = \int_X f(x)\mu(dy, x)$ where $\mu(,x)$ is a measure of $(X, \mathcal{B})$ for all $x \in X$, and each function X $\mu(B, x)$ is measurable for all $B \in \mathcal{B}$. This is only a mild restriction.

Note that a definition by application Riesz if X is locally compact Hausdorff and $\mathcal{B}_X$ is the Borel-sigma algebra. Suppose $R(1) = 1$, then the following representation of $\mathbb{P}$ on $(\text{Sol}(X), \text{cylindersets}, \mathbb{P})$ are equivalent: The following are equivalent

(i) $\int_{\text{Sol}}(f \circ \pi_0)(g \circ \pi_n)d\mathbb{P} = \int_X f(x)(R^n g)(x)d\lambda(x);$
(ii) $\text{Prob}_{\text{w.r.t } \mathbb{P}}(\pi_0 = x, \pi_1 \in B_1, \ldots, \pi_k \in B_k) = \int_{B_1} \int_{B_2} \cdots \int_{B_k} \mu(dy_1|x)\mu(dy_2|y_1) \cdots \mu(dy_k|y_{k-1})$ for all k and for all $B_i \in \mathcal{B}_X;$
(iii) $\text{Prob}_{\text{w.r.t } \mathbb{P}}(\pi_0 = x, \pi_1 \in B_1, \cdots, \pi_k \in B_k) = R(\chi_{B_1} R(\chi_{B_2} R \cdots R (\chi_{B_k}) \cdots (x)) = \mathbb{E}_x(\cdots) = \int_{\text{Sol}} \cdots d\mathbb{P}_x.$

In the case R, $R1 = 1$. But if not, then pick h such that $Rh = h$, and set $R'(f) = \frac{R(fh)}{h}$ with $R'(1) = 1$.

$$\text{Prob}(\pi_0 = x, \pi_1 = y_1, \ldots, \pi_k = y_k)$$

$$\frac{1}{N^k} W(y_1)W(y_2) \cdots W(y_k)$$

$$Pr(x \to y_1)Pr(y_1 \to y_2) \cdots Pr(y_{k-1} \to y_k)$$

$$Rf(x) = \int_X f(y)\mu(dy|x) \quad \text{represent } R \text{ as an....}$$

$$\mu(B|x) := R(\chi_B)(x), \quad \forall B \in \mathcal{B}$$
$$= \text{Prob}(\pi_1 \in B | \pi_0 = x).$$

More generally:

$$\text{Prob}(\pi_0 = x, \ \pi_1 \in B_1, \dots, \pi_k \in B_k)$$
$$= \int_{B_1} \int_{B_2} \cdots \int_{B_k} \mu(dy_1|x)\mu(dy_2|y_1)\cdots\mu(dy_k|y_{k-1})$$
$$= R(\chi_{B_1} R(\chi_{B_2} R \cdots R(\chi_{B_k}) \cdots (x)).$$

Same manner prop of $\{\mu(--|x)\}_{x \in X}$.

Lemma D.5. *If $B \in \mathcal{B}_X$, then*

$$\int_X \mu(B|x)d\lambda(x) = \int_B W(x)d\lambda(x), \quad \text{where } W = \frac{d\lambda(R)}{d\lambda}.$$

Proof. Let $\{\mu(B|x)\}_{x \in X}$ be a Markov process by $x \in X$ and (X, B) is a fixed measure space and let $\mathbb{P}$ be the corresponding path space measure $\mathbb{P}(\pi_0 = x, \ \pi_1 \in B_1, \dots, \pi_k \in B_k)$. Let

$$\sigma \in \text{End}(XB) = \int_{B_1}\int_{B_2}\cdots\int_{B_k}\mu(dy_1|x)\mu(dy_2|y_1)\cdots\mu(dy_k|y_{k-1}),$$

then $\mathbb{P}$ is ... as equation σ-solenoid $\text{Sol}_\sigma(X)$ if and only if

$$\mathbb{P}(\pi_{k-1} \in B \cap \sigma^{-1}(A)|\pi_k = x) = \chi_A(x)\mathbb{P}(\pi_k \in B|\pi_k = x)$$

if and only if $\text{supp}(\mathbb{P}) \subset \text{Sol}_\sigma(X)$. □

Lemma D.6. *Suppose R has a representation*

$$R(\chi_B)(x) = \mu(B|x), \quad B \in \mathcal{B}_X, \quad x \in X. \tag{D.2}$$

The following are equivalent: as described, i.e., $(Rf)(x) = \int_X f(x)\mu(dy|x)$; then

$$R[(f \circ \sigma)g](x) = f(x)R(g)(x), \quad \forall x, \quad \forall f, g. \tag{D.3}$$

If and only if

$$\mu(\sigma^{-1}(A) \cap B|x) = \chi_A(x)\mu(B|x) \quad \forall A, B \in \mathcal{B} \ \forall x \in X. \tag{D.4}$$

Notation: Let $\{\mu(\cdot|x)\}_{x \in X}$ be the family of measures on (x, B); define R as in (D.2); and set $\mu_x(\cdot) := \mu(\cdot|x)$ then (D.3) if and only if (conditional measures):

$$\mu_x(\sigma^{-1}(A)|B) = \chi_A(x), \quad \text{where } \mu_x(\cdot|B)$$

denote the conditional probability, i.e.,

$$\mu_x(\sigma^{-1}(A)|B) = \begin{cases} 1 & \text{if } x \in A \\ 0 & \text{if } x \notin A \end{cases}$$

$$\mu_x(\sigma^{-1}(A))\mu_x(B|\sigma^{-1}(A)) =$$
$$\mu_x(\sigma^{-1}(A) \cap B) = \chi_A(x)\mu_x(B)$$
$$\mu_x(\sigma^{-1}(A)|B) = \chi_A(x)$$
$$\mu_n(\Delta) = \mathbb{P}(\pi_n^{-1}(\Delta)).$$

Proof.

$$\chi_A(x)\mu_x(B|\sigma^{-1}(A)) = \mu(B)\mu_x(\sigma^{-1}(A)|B)$$
$$\chi_A(x)\mu_x(B|\sigma^{-1}(A)) = \chi_A(x)\mu_x(B) = \mu(B)\mu_x(\sigma^{-1}(A)|B)\mu_x(B) \qquad \square$$

(i) (LP) $\frac{m_0(0)}{\sqrt{N}} = 1$ low-pass. $W = |r_0|$ $(Rf)(x) = \frac{1}{N}\sum_{\sigma(y)=x}(Wf)(y)$

(ii) $R(h) = h$

$$\prod_{k=1}^{\infty} \frac{m_0(t/N^k)}{\sqrt{N}} \to |\hat{\varphi}(t)|^2$$

$$\prod_{k=1}^{\infty} \frac{m_0(t/N^k)}{\sqrt{N}} \in L^2(\mathbb{R}).$$

Theorem D.7. $(X, \sigma, R, \lambda) \to \mathrm{Sol}_\sigma(X)$. *We have*

(i) *$\exists!$ $\mathbb{P}$ such that $dist(\pi_k) = \mu_k$, $\int_X f d\mu_R = \int_X R^k(f)d\lambda$, and*
(ii) *$\mathbb{P}$ has the property: $\frac{d\mathbb{P}\cdot\hat{\sigma}}{d\mathbb{P}} = w$*

   Given

$$X \xrightarrow{f} \mathbb{R} \quad \text{we get}$$

$$X \xrightarrow{f \circ \pi_n} \mathrm{Sol}$$

$$V_n f := f \circ \pi_n, \quad \text{where}$$

$$\mu_n := dist(\pi_n)$$

$$\mu_n := \mathbb{P} \circ \pi_n^{-1}$$

$$V_n : L^2(X, \mu_n) \xrightarrow{\text{isometry}} \mathrm{Sol}(X)$$

$$V_n : f \to f \circ \pi_n.$$

Here, "dist" is short for distribution.

The proof follows from the above discussion.

$\mathcal{H} : (\Omega, \mathcal{F}, \mathbb{P})$. Let λ be on X and $\lambda R << \lambda$ Radon–Nikodym on W. Let U be on $F \in L^2(\Omega, \mathbb{P})$, $R(f)d\lambda = \int fwd\lambda$, $m_0 = \sqrt{W \circ \pi_0}$, $UF = \sqrt{W \circ \pi_0} F \circ \hat{\sigma}$ [JS18].

Theorem D.8. *Let $L^2(\Omega, \mathbb{P})$, $\mathbf{1}$, U, ρ. Then there exists a Hilbert space $\mathcal{H}$, a representation ρ of $L^2(\Omega, \mathbb{P})$ on $\mathcal{H}$, a unitary operator U on $\mathcal{H}$ and a vector φ in $\mathcal{H}$ such that*

$$\rho(f) = m(f \circ \pi_0)$$

$$f \in L^\infty(X) \quad F \to (f \circ \pi_0)F$$

$(\mathcal{H}, \varphi, U, \rho) \to$ *path space measure.*

(i) (Covariance)

$$U\rho(f) = \rho(f \circ r)U \quad f \in L^\infty(X). \tag{D.5}$$

(ii) (Scaling equation)

$$F \to (f \circ \pi_0)F \to (f \circ \sigma) \circ \pi_0 \sqrt{W \circ \pi_0} F \circ \hat{\sigma}. \tag{D.6}$$

(iii) (Orthogonality)

$$\int_\Omega \rho(f) 1 d\mathbb{P} = \int (f \circ \pi_0) d\lambda = \int f d\lambda. \tag{D.7}$$

(iv) (The inverse to the operator U)

$$U^{-1} F = \frac{1}{W \circ \pi_1} F \circ \hat{\sigma}^{-1} \quad \pi_n(w) = x_n. \tag{D.8}$$

Example D.9.

$$UF = \sqrt{W \circ \pi_0} F \circ \hat{\sigma} \quad \text{where } \sigma : X \to X$$

$$U^* = U^{-1}$$

$$U = \frac{1}{\sqrt{2}} f\left(\frac{x}{2}\right)$$

$$U\hat{\varphi} = m\hat{\varphi} \quad \text{where } \hat{\varphi} \in L^2(\mathbb{R}).$$

We are interested in finding the filter function analogous to Dutkay–Jorgensen Haar–Cantor filter. $m_0 = \frac{1+z}{\sqrt{2}}$.

An attempt to find filter functions for Julia set: Let $m_i : X_c \to \mathbb{C}$. For Julia set, $r(z) = z^N$ on $\mathbb{T}$. μ is Haar measure on $\mathbb{T}$. Strong invariance of μ with respect to $r(z)$.

$$\int \frac{1}{\#r^{-1}(z)} \sum_{r(w)=z} f(w) dz = \int f(z) d\mu(z).$$

By Brolin's theorem, there exists a unique μ strongly invariant for r [JS18].

Quadrature mirror filter

$$\frac{1}{\#r^{-1}(z)} \sum_{r(w)=1} |m_0(w)|^2 = 1 \quad z \in J_c.$$

We want to find nice m_0 and m_1.

$$w^2 + c = z \Rightarrow w_\pm = \pm\sqrt{z-c} \text{ where } w_1 = \sqrt{z-c} \text{ and } w_2 = -\sqrt{z-c}.$$

$$\frac{1}{2}(|m_0(w_1)|^2 + |m_0(w_2)|^2) = 1$$

$$\frac{1}{2}(|m_1(w_1)|^2 + |m_1(w_2)|^2) = 1$$

$$\frac{1}{2}(m_0(w_1)m_1(w_1) + m_0(w_2)m_1(w_2)) = 1.$$

We are interested in solving the following matrix over polynomial

$$\frac{1}{\sqrt{2}} \begin{bmatrix} m_0(w_1) & m_0(w_2) \\ m_1(w_1) & m_1(w_2) \end{bmatrix},$$

where it is unitary and $m_0 = 1$.

Also, we would like to find m_1, high-pass filter

$$\frac{1}{2}(m_1(\sqrt{z-c}) + m_1(-\sqrt{z-c})) = 0$$

$$\frac{1}{2}(|m_1(\sqrt{z-c})|^2 + |m_1(-\sqrt{z-c})|^2) = 1,$$

where $m_1(w) = -m(-w)$.

References

[Abd08] Fatma Abdelkefi, Performance of sigma-delta quantizations in finite frames, *IEEE Trans. Inform. Theory* **54** (2008), no. 11, 5087–5101. MR MR2589884 (2010g:94034).

[AK06a] William Arveson and Richard V. Kadison, Diagonals of self-adjoint operators, Operator theory, operator algebras, and applications, *Contemp. Math.*, vol. 414, Amer. Math. Soc., Providence, RI, 2006, pp. 247–263. MR MR2277215.

[AK06b] Gilles Aubert and Pierre Kornprobst, *Mathematical Problems in Image Processing*, Springer, New York, 2006.

[Akh65] N. I. Akhiezer, *The Classical Moment Problem and Some Related Questions in Analysis*, Translated by N. Kemmer, Hafner Publishing Co., New York, 1965. MR MR0184042 (32 #1518).

[AMWW16] Habib Ammari, Stéphane Mallat, Irène Waldspurger, and Han Wang, Wavelet methods for shape perception in electro-sensing, Imaging, multi-scale and high contrast partial differential equations, *Contemp. Math.*, vol. 660, Amer. Math. Soc., Providence, RI, 2016, pp. 1–21. MR 3485667.

[Arv07] William Arveson, Diagonals of normal operators with finite spectrum, *Proc. Natl. Acad. Sci. USA* **104** (2007), no. 4, 1152–1158 (electronic). MR MR2303566.

[AS99] E. Aboufadel and S. Schilicker, *Discovering Wavelets*, Wiley-Interscience, John Wiley & Sons, Inc., New York, 1999.

[Ash90] Robert B. Ash, *Information Theory*, Dover Publications Inc., New York, 1990, Corrected reprint of the 1965 original. MR MR1088248 (91k:94001).

[AW12] Arash A. Amini and Martin J. Wainwright, Sampled forms of functional PCA in reproducing kernel Hilbert spaces, *Ann. Statist.* **40** (2012), no. 5, 2483–2510. MR 3097610.

[Bal00] Viviane Baladi, *Positive Transfer Operators and Decay of Correlations*, Advanced Series in Nonlinear Dynamics, vol. 16, World Scientific Publishing Co. Inc., River Edge, New Jersey, 2000. MR 1793194 (2001k:37035).

[BCW90] T. C. Bell, J. G. Cleary, and I. H. Witten, *Text Compression*, Prentice Hall Advanced Reference Series: Computer Science, Prentice Hall, Hoboken, New Jersey, USA, 1990.

[Ben48] W. R. Bennett, Spectra of quantized signals, *Bell System Tech. J.* **27** (1948), 446–472. MR MR0026287 (10,133f).

[BJ99] Ola Bratteli and Palle E. T. Jorgensen, Iterated function systems and permutation representations of the Cuntz algebra, *Mem. Amer. Math. Soc.* **139** (1999), no. 663, x+89. MR 1469149.

[BJ02a] Ola Bratteli and Palle Jorgensen, *Wavelets through a Looking Glass: The World of the Spectrum*, Applied and Numerical Harmonic Analysis, Birkhäuser Boston Inc., Boston, MA, 2002, MR MR1913212 (2003i:42001).

[BJ02b] Ola Bratteli and Palle E. T. Jorgensen, Wavelet filters and infinite-dimensional unitary groups, Wavelet analysis and applications (Guangzhou, 1999), *AMS/IP Stud. Adv. Math.*, vol. 25, Amer. Math. Soc., Providence, RI, 2002, pp. 35–65. MR MR1887500 (2003e:94015).

[BJMP05] Lawrence Baggett, Palle Jorgensen, Kathy Merrill, and Judith Packer, A non-MRA C^r frame wavelet with rapid decay, *Acta Appl. Math.*, Springer, New York (2005).

[BLM06] Bredies, K., Lorenz, D.A., Maass, P. (2006). An Optimal Control Problem in Medical Image Processing. In: Ceragioli, F., Dontchev, A., Furuta, H., Marti, K., Pandolfi, L. (eds.), Systems, Control, Modeling and Optimization. CSMO 2005. IFIP International Federation for Information Processing, vol. 202. Springer, Boston, MA. https://doi.org/10.1007/0-387-33882-9_23

[BOT08] John J. Benedetto, Onur Oktay, and Aram Tangboondouangjit, Complex sigma-delta quantization algorithms for finite frames, Radon transforms, geometry, and wavelets, *Contemp. Math.*, vol. 464, Amer. Math. Soc., Providence, RI, 2008, pp. 27–49. MR MR2440128 (2009m:94034).

[BR91] Chris Brislawn and I. G. Rosen, Wavelet based approximation in the optimal control of distributed parameter systems, *Numer. Funct. Anal. Optim.* **12** (1991), no. 1–2, 33–77. MR MR1125045 (92g:49029).

[Bra06] Mark Braverman, Parabolic Julia sets are polynomial time computable, *Nonlinearity* **19** (2006), no. 6, 1383–1401. MR 2230004.

[Bri10a] Christopher M. Brislawn, Group lifting structures for multirate filter banks I: uniqueness of lifting factorizations, *IEEE Trans. Signal Process.* **58** (2010), no. 4, 2068–2077. MR 2680914.

[Bri10b] Christopher M. Brislawn, Group lifting structures for multirate filter banks II: linear phase filter banks, *IEEE Trans. Signal Process.* **58** (2010), no. 4, 2078–2087. MR 2680915.

[Bri13a] Christopher M. Brislawn, Group-theoretic structure of linear phase multirate filter banks, *IEEE Trans. Inform. Theory* **59** (2013), no. 9, 5842–5859. MR 3096961.

[Bri13b]   Christopher M. Brislawn, On the group-theoretic structure of lifted filter banks, *Excursions in harmonic analysis.* vol. 2, Appl. Numer. Harmon. Anal., Birkhäuser/Springer, New York, 2013, pp. 113–135. MR 3050316.

[BWP03]   C. M. Brislawn, B. E. Wohlberg, and A. G. Percus, Resolution scalability for arbitrary wavelettransforms in the jpeg-2000 standard, *Visual Commun. & Image Process., Ser. Proc. SPIE* **5150** (2003), no. 1, 774–784.

[BY06]   M. Braverman and M. Yampolsky, Non-computable Julia sets, *J. Amer. Math. Soc.* **19** (2006), no. 3, 551–578. MR 2220099.

[CC08]   X. X. Chen and Y. Y. Chen, Self-lifting scheme: new approach for generating and factoring wavelet filter bank, *IET Signal Process.* **2** (2008), no. 4, 405–414. MR MR2522956.

[CCM16]   Xiuyuan Cheng, Xu Chen, and Stéphane Mallat, Deep Haar scattering networks, *Inf. Inference* **5** (2016), no. 2, 105–133. MR 3516855.

[CD04]   Emmanuel J. Candès and David L. Donoho, New tight frames of curvelets and optimal representations of objects with piecewise C^2 singularities, *Comm. Pure Appl. Math.* **57** (2004), no. 2, 219–266. MR 2012649.

[CDDD03]   Albert Cohen, Wolfgang Dahmen, Ingrid Daubechies, and Ronald DeVore, Harmonic analysis of the space BV, *Rev. Mat. Iberoamericana* **19** (2003), no. 1, 235–263. MR 1993422.

[CG93]   Lennart Carleson and Theodore W. Gamelin, *Complex Dynamics*, Universitext: Tracts in Mathematics, Springer-Verlag, New York, 1993. MR 1230383.

[CGV99]   A. Cohen, K. Gröchenig, and L. F. Villemoes, Regularity of multivariate refinable functions, *Constr. Approx.* **15** (1999), no. 2, 241–255. MR 1668921.

[Chr03]   Ole Christensen, *An Introduction to Frames and Riesz Bases*, Applied and Numerical Harmonic Analysis, Birkhäuser Boston Inc., Boston, MA, 2003. MR MR1946982 (2003k:42001).

[CK04]   Peter G. Casazza and Gitta Kutyniok, Frames of subspaces, *Wavelets, frames and operator theory, Contemp. Math.*, vol. 345, Amer. Math. Soc., Providence, RI, 2004, pp. 87–113. MR MR2066823 (2005e:42090).

[CKL08]   Peter G. Casazza, Gitta Kutyniok, and Shidong Li, Fusion frames and distributed processing, *Appl. Comput. Harmon. Anal.* **25** (2008), no. 1, 114–132. MR 2419707 (2009d:42094).

[Coh03a]   A. Cohen, Multiscale adaptive processing for evolution equations, *Numerical Mathematics and Advanced Applications*, Springer Italia, Milan, 2003, pp. 605–629. MR 2360759.

[Coh03b]   Albert Cohen, *Numerical Analysis of Wavelet Methods*, Studies in Mathematics and its Applications, vol. 32, North-Holland Publishing Co., Amsterdam, 2003. MR 1990555.

[Dau92]   Ingrid Daubechies, *Ten Lectures on Wavelets*, CBMS-NSF Regional Conference Series in Applied Mathematics, vol. 61, Society for Industrial and Applied Mathematics (SIAM), Philadelphia, PA, 1992. MR 1162107 (93e:42045).

[Dau93]   Ingrid Daubechies, Wavelet transforms and orthonormal wavelet bases, *Proc. Sympos. Appl. Math.*, 1993.

[DDLY00]   David L. Donoho, Nira Dyn, David Levin, and Thomas P. Y. Yu, Smooth multiwavelet duals of Alpert bases by moment-interpolating refinement, *Appl. Comput. Harmon. Anal.* **9** (2000), no. 2, 166–203. MR 1777125.

[Dir47]   P. A. M. Dirac, *The Principles of Quantum Mechanics*, 3rd edition, Clarendon Press, Oxford, 1947, MR MR0023198 (9,319d).

[DJ05]   Dorin Ervin Dutkay and Palle E. T. Jorgensen, Wavelet constructions in non-linear dynamics, *Electron. Res. Announc. Amer. Math. Soc.* **11** (2005), 21–33. MR 2122446.

[DJ06a]   Dorin E. Dutkay and Palle E. T. Jorgensen, Wavelets on fractals, *Rev. Mat. Iberoamericana* **22** (2006), no. 1, 131–180. MR 2268116.

[DJ06b]   Dorin Ervin Dutkay and Palle E. T. Jorgensen, Hilbert spaces built on a similarity and on dynamical renormalization, *J. Math. Phys.* **47** (2006), no. 5, 053504, 20. MR 2239365.

[DJ06c]   Dorin Ervin Dutkay and Palle E. T. Jorgensen, Iterated function systems, Ruelle operators, and invariant projective measures, *Math. Comp.* **75** (2006), no. 256, 1931–1970. MR 2240643.

[DJ07]   Dorin Ervin Dutkay and Palle E. T. Jorgensen, Disintegration of projective measures, *Proc. Amer. Math. Soc.* **135** (2007), no. 1, 169–179 (electronic). MR MR2280185.

[DL92]   Ingrid Daubechies and Jeffrey C. Lagarias, Two-scale difference equations. II. Local regularity, infinite products of matrices and fractals, *SIAM J. Math. Anal.* SIAM, Philadelphia, PA, USA, **23** (1992), no. 4, 1031–1079. MR 1166574.

[DL06]   Robert L. Devaney and Daniel M. Look, A criterion for Sierpinski curve Julia sets, *Topology Proc.* **30** (2006), no. 1, 163–179, Spring Topology and Dynamical Systems Conference. MR MR2280665.

[dLMSS56]   K. de Leeuw, E. F. Moore, C. E. Shannon, and N. Shapiro, Computability by probabilistic machines, *Automata Studies*, Annals of Mathematics Studies, no. 34, Princeton University Press, Princeton, New Jersey, 1956, pp. 183–212. MR 0079550.

[Don00]   David L. Donoho, Orthonormal ridgelets and linear singularities, *SIAM J. Math. Anal.* **31** (2000), no. 5, 1062–1099. MR 1759199.

[Dou98]   R. G. Douglas, *Banach Algebra Techniques in Operator Theory*, 2nd edition, Springer, New York, 1998.

[DPS14]   Dorin Ervin Dutkay, Gabriel Picioroaga, and Myung-Sin Song, Orthonormal bases generated by Cuntz algebras, *J. Math. Anal. Appl.* **409** (2014), no. 2, 1128–1139. MR 3103223.

[DR07]   Dorin Ervin Dutkay and Kjetil Røysland, The algebra of harmonic functions for a matrix-valued transfer operator, *J. Funct. Anal.* **252** (2007), no. 2, 734–762. MR 2360935 (2008m:42056).

[DR08]   Dorin Ervin Dutkay and Kjetil Røysland, Covariant representations for matrix-valued transfer operators, *Integral Equations Operator Theory* **62** (2008), no. 3, 383–410. MR 2461126 (2010k:42066).

[DRS07]   Robert L. Devaney, Mónica Moreno Rocha, and Stefan Siegmund, Rational maps with generalized Sierpinski gasket Julia sets, *Topology Appl.* **154** (2007), no. 1, 11–27. MR MR2271770.

[DS97]   G. M. D'Ariano and M. F. Sacchi, Optical von Neumann measurement, *Phys. Lett. A* **231** (1997), no. 5–6, 325–330. MR 1469159 (98c:81019).

[DS98]   Ingrid Daubechies and Wim Sweldens, Factoring wavelet transforms into lifting steps, *J. Fourier Anal. Appl.* **4** (1998), no. 3, 247–269. MR MR1650921 (99g:42040).

[Dut04]   Dorin Ervin Dutkay, The spectrum of the wavelet Galerkin operator, *Integral Equations Operator Theory*, Birkhäuser Verlag, Springer Nature Switzerland AG, **50** (2004), no. 4, 477–487. MR 2105959.

[DVDD98]   David L. Donoho, Martin Vetterli, R. A. DeVore, and Ingrid Daubechies, Data compression and harmonic analysis, *IEEE Trans. Inform. Theory* **44** (1998), no. 6, 2435–2476, Information Theory: 1948–1998. MR MR1658775 (99i:94028).

[Eld03]   Yonina C. Eldar, von Neumann measurement is optimal for detecting linearly independent mixed quantum states, *Phys. Rev. A (3)* **68** (2003), no. 5, 052303, 4. MR MR2026802 (2004k:81047).

[GM85]   A. Grossmann and J. Morlet, Decomposition of functions into wavelets of constant shape, and related transforms, *Mathematics + Physics.* vol. 1, World Sci. Publishing, Singapore, 1985, pp. 135–165. MR 849345.

[GvN48]   H. H. Goldstine and J. von Neumann, *Planning and Coding of Problems for an Electronic Computing Instrument*, Report on the Mathematical and Logical Aspects of an Electronic Computing Instrument, Part II, Vol. 3, The Institute for Advanced Study, Princeton, New Jersey, iii–23, 1948.

[GW02]   R. C. Gonzalez and R. E. Woods, *Digital Image Processing*, 2nd edition, Prentice Hall, Hobeken, New Jersey, USA, 2002.

[GWE04]   R. C. Gonzalez, R. E. Woods, and S. L. Eddins, *Digital Image Processing Using Matlab*, Prentice Hall, Hobeken, New Jersey, USA, 2004.

[Haa10]   Alfred Haar, Zur Theorie der orthogonalen Funktionensysteme, *Math. Ann.* **69** (1910), no. 3, 331–371. MR 1511592.

[HCXH08]   Yumin He, Xuefeng Chen, Jiawei Xiang, and Zhengjia He, Multiresolution analysis for finite element method using interpolating wavelet and lifting scheme, *Comm. Numer. Methods Engrg.* **24** (2008), no. 11, 1045–1066. MR MR2474670.

[HKMMT89]   M. Holschneider, R. Kronland-Martinet, J. Morlet, and Ph. Tchamitchian, A real-time algorithm for signal analysis with the help of the wavelet transform, *Wavelets* (Marseille, 1987), Inverse

Probl. Theoret. Imaging, Springer, Berlin, 1989, pp. 286–297. MR 1010915.

[Hut81]   John E. Hutchinson, Fractals and self-similarity, *Indiana Univ. Math. J.* **30** (1981), no. 5, 713–747. MR MR625600 (82h:49026).

[HW06]   Christopher Heil and David F. Walnut (eds.), *Fundamental Papers in Wavelet Theory*, Princeton University Press, Princeton, New Jersey, 2006. MR MR2229251.

[Iwa49]   Kenkichi Iwasawa, On some types of topological groups, *Ann. of Math. (2)* **50** (1949), 507–558. MR MR0029911 (10,679a).

[JlCH01] A. Jensen and A. la Cour-Harbo, *Ripples in Mathematics: The Discrete Wavelet Transform*, Springer-Verlag, Berlin, 2001. MR MR1839000 (2002e:94029).

[JMR01]   Stéphane Jaffard, Yves Meyer, and Robert D. Ryan, *Wavelets*, revised edition, Society for Industrial and Applied Mathematics (SIAM), Philadelphia, PA, 2001, Tools for Science & Technology. MR 1827998.

[Jor01a] Palle E. T. Jorgensen, Minimality of the data in wavelet filters, *Adv. Math.* **159** (2001), no. 2, 143–228, with an appendix by Brian Treadway. MR 1825057.

[Jor01b] Palle E. T. Jorgensen, Representations of Cuntz algebras, loop groups and wavelets, *XIIIth International Congress on Mathematical Physics (London, 2000)*, Int. Press, Boston, MA, 2001, pp. 327–332. MR 1883323 (2002k:46144).

[Jor03]   Palle E. T. Jorgensen, Matrix factorizations, algorithms, wavelets, *Notices Amer. Math. Soc.* **50** (2003).

[Jor05]   P. E. T. Jorgensen, *Analysis and Probability Wavelets, Fractals, and Dynamics*, Springer, New York, 2005.

[Jor06a] Palle E. T. Jorgensen, *Analysis and Probability: Wavelets, Signals, Fractals*, Graduate Texts in Mathematics, vol. 234, Springer, New York, 2006.

[Jor06b] Palle E. T. Jorgensen, Certain representations of the Cuntz relations, and a question on wavelets decompositions, vol. 414, Amer. Math. Soc., Providence, RI, 2006, pp. 165–188. MR 2277210.

[JS07]   Palle E. T. Jorgensen and Myung-Sin Song, Entropy encoding, Hilbert space, and Karhunen-Loève transforms, *J. Math. Phys.* **48** (2007), no. 10, 103503.

[JS09]   Palle E. T. Jorgensen and Myung-Sin Song, Analysis of fractals, image compression, entropy encoding, Karhunen-Loève transforms, *Acta Appl. Math.* **108** (2009), no. 3, 489–508. MR MR2563494.

[JS10]   P. E. T. Jorgensen and M. S. Song, Matrix factorization and lifting, *Sampling Theory in Signal and Image Processing*, http://stsip.org, Sample Publishing, USA, (2010), no. 9, 167–197. MR 2814346.

[JS14a]   P. E. T. Jorgensen and M. S. Song, *Comparison of discrete and continuous wavelet transforms*, Springer Encyclopedia of Complexity and Systems Science, Springer, New York (2014).

[JS14b]   P. E. T. Jorgensen and M. S. Song, Filters and matrix factorization, *Sampling Theory in Signal and Image Processing*, http://stsip.org, Sample Publishing, USA (2014), no. 14, 171–197.

[JS18]   P. E. T. Jorgensen and M. S. Song, Markov chains and generalized wavelet multiresolutions, *Journal of Analysis*, Springer, New York (2018), no. 26, 259–283.

[JT21]   Palle Jorgensen and James Tian, Infinite-Dimensional Analysis— Operators in Hilbert Space; Stochastic Calculus via Representations, and Duality Theory, World Scientific Publishing Co. Pte. Ltd., Hackensack, New Jersey, 2021. MR 4274591.

[Kar46]   Kari Karhunen, *Zur spektraltheorie stochastischer prozesse*, Annales Academiae scientiarum Fennicae. Series A.1, Mathematica-physica, Acad. Sci. Fennica (Finnish Academy of Science and Letters), Finland, 1946, p. 7. MR 23012.

[Kar52]   Kari Karhunen, *Über ein extrapolationsproblem in dem hilbertschen raum*, Den 11te Skandinaviske Matematikerkongress, Trondheim (1949) **18** (1952), 35–41.

[Kei04]   F. Keinert, *Wavelets and Multiwavelets*, Chapmand & Hall/CRC, UK, 2004.

[KJST]   Sooran Kang, Palle E. T. Jorgensen, Myung-Sin Song, and Feng Tian, *An Infinite Dimensional Analysis of Kernel Principal Components*, arXiv:1906.06451.

[Kol77]   A. Kolmogoroff, *Grundbegriffe der Wahrschein- lichkeitsrechnung*, Springer-Verlag, Berlin, 1977, Reprint of the 1933 original. MR MR0494348 (58 #13242).

[KR97]   Richard V. Kadison and John R. Ringrose, *Fundamentals of the Theory of Operator Algebras*. vol. I. Graduate Studies in Mathematics, vol. 15, American Mathematical Society, Providence, RI, 1997, Elementary theory, Reprint of the 1983 original. MR MR1468229 (98f:46001a).

[Kri05]   D. W. Kribs, A quantum computing primer for operator theorists, *Linear Algebra Appl.* **400** (2005), 147–167.

[Law99]   Wayne M. Lawton, *Conjugate quadrature filters*, Advances in Wavelets (Hong Kong, 1997), Springer, Singapore, 1999, pp. 103–119. MR MR1688766 (2000g:42043).

[Law00]   Wayne Lawton, Infinite convolution products and refinable distributions on Lie groups, *Trans. Amer. Math. Soc.* **352** (2000), no. 6, 2913–2936. MR MR1638258 (2000j:43002).

[Law02]   Wayne Lawton, Global analysis of wavelet methods for Euler's equation, *Mat. Model.* **14** (2002), no. 5, 75–88, Second International Conference OFEA'2001 "Optimization of Finite Element Approximation, Splines and Wavelets" (Russian) (St. Petersburg, 2001). MR MR1937337 (2003k:37143).

[Law04]   Wayne M. Lawton, Hermite interpolation in loop groups and conjugate quadrature filter approximation, *Acta Appl. Math.* **84** (2004), no. 3, 315–349. MR MR2117661 (2005i:41025).

[LHR09]   B. W. K. Ling, C. Y. F. Ho, and J. D. Reiss, *Control of Sigma Delta Modulators via Fuzzy Impulsive Approach*, Control of Chaos in Nonlinear Circuits and Systems, World Sci. Ser. Nonlinear Sci. Ser. A Monogr. Treatises, vol. 64, World Sci. Publ., Hackensack, New Jersey, 2009, pp. 245–270. MR MR2523485.

[Liu06]   Feng Liu, *Diffusion Filtering in Image Processing Based on Wavelet Transform*, Sci. China Ser. F and Springer, New York, **49** (2006), no. 4, 494–503. MR 2265315.

[Loe55]   Michel Loeve, *Probability Theory: Foundations, Random Sequences*, D. Van Nostrand Company, Inc., Toronto-New York-London, xv+515 pp. 1955.

[LPY10]   M. Lammers, A. M. Powell, and Özgür Yılmaz, Alternative dual frames for digital-to-analog conversion in sigma-delta quantization, *Adv. Comput. Math.* **32** (2010), no. 1, 73–102. MR MR2574568.

[Mac01]   D. Mackenzie, *Wavelets Seeing the Forest — and the Trees*, Beyond Discovery (Dec. 2001), http://www.nasonline.org/publications/beyond-discovery/wavelets.pdf.

[Mal09]   Stéphane Mallat, *A Wavelet Tour of Signal Processing: The Sparse Way*, 3rd edition, Elsevier/Academic Press, Amsterdam, 2009, The sparse way, With contributions from Gabriel Peyré. MR 2479996.

[Mar82]   D. Marr, *Vision*, W. H. Freeman and Company, New York, 1982.

[Mar14]   S. Marsland, *Machine Learning: An Algorithmic Perspective*, Chapman and Hall/CRC, Boca Raton, FL, 2014.

[Mey00a] Yves Meyer, Le traitement du signal et l'analyse mathématique, *Ann. Inst. Fourier (Grenoble)* **50** (2000), no. 2, 593–632. MR 1775362.

[Mey00b] Yves Meyer, Wavelets and functions with bounded variation from image processing to pure mathematics, *Atti Accad. Naz. Lincei Cl. Sci. Fis. Mat. Natur. Rend. Lincei (9) Mat. Appl.* (2000), no. Special Issue, 77–105, Mathematics towards the third millennium (Rome, 1999). MR 1845666.

[Mil04]   John Milnor, Pasting together Julia sets: A worked out example of mating, *Experiment. Math.* **13** (2004), no. 1, 55–92. MR MR2065568 (2005c:37087).

[MMOP02] M. Misitti, Y. Misitti, G. Oppenheim, and J. Poggi, *Wavelet Toolbox for Use with Matlab, Version 2*, The MathWorks, Natick, MA, USA, 2002.

[MP05]   M. Mudrova and A. Prochazka, Principal component analysis in image processing, *Proceedings of the MATLAB Technical Computing Conference*, Prague (2005).

[PBK07]   Sangwoong Park, Joonwoo Bae, and Younghun Kwon, Wavelet quantum search algorithm with partial information, *Chaos Solitons Fractals* **32** (2007), no. 4, 1371–1374. MR 2286303.

[PTVF92] W. H. Press, S. A. Teukolsky, W. T. Vetterling, and B. P. Flannery, *Numerical Recipes in C: The Art of Scientific Computing*, 2nd edition, Cambridge University Press, 1992.

[PZ04]   C. L. Petersen and S. Zakeri, On the Julia set of a typical quadratic polynomial with a Siegel disk, *Ann. of Math. (2)* **159** (2004), no. 1, 1–52. MR MR2051390 (2005c:37085).

[Reb20]   Jason Rebello, *Principal Component Analysis (PCA) (https:// www.mathworks.com/matlabcentral/fileexchange/42847-princip al-component-analysis-pca)*, MATLAB Central File Exchange (2020).

[Res05]   Wolfram Research, *Wavelet Explorer Documentation*, Wolfram Research, Inc., Urbana, IL, USA, 2005.

[Rue69]   David Ruelle, *Statistical Mechanics: RIgorous Results*, W. A. Benjamin, Inc., New York-Amsterdam, 1969. MR 0289084.

[Rue02]   David Ruelle, *Dynamical Zeta Functions and Transfer Operators*, Notices Amer. Math. Soc. **49** (2002), no. 8, 887–895. MR 1920859.

[RW98]   H. L. Resnikoff and R. O. Jr. Wells, *Wavelet Analysis*, Springer, New York, 1998.

[SBT03]   Peng-Lang Shui, Zheng Bao, and Yuan Yan Tang, Three-band biorthogonal interpolating complex wavelets with stopband suppression via lifting scheme, *IEEE Trans. Signal Process.* **51** (2003), no. 5, 1293–1305. MR MR2049909.

[SCE01]   A. Skodras, C. Christopoulos, and T. Ebrahimi, Jpeg 2000 still image compression standard, *IEEE Signal Processing Magazine* **18** (Sept. 2001), 36–58.

[Sha57]   Claude E. Shannon, Certain Results in Coding Theory for Noisy Channels, *Information and Control* **1** (1957), 6–25. MR 92707.

[Sha60]   Claude E. Shannon, *Coding Theorems for a Discrete Source with a Fidelity Criterion*, Information and Decision Processes, McGraw-Hill, New York, 1960, pp. 93–126. MR MR0122612 (22 #13335).

[Sha93a]   Claude Elwood Shannon, IEEE Press, New York, 1993, Collected papers, Edited by N. J. A. Sloane and Aaron D. Wyner, With a profile of Shannon by Anthony Liversidge. MR 1216351.

[Sha93b]   J. Shapiro, Embedded image coding using zerotrees of wavelet coefficients, *IEEE Trans. Signal Processing* **41** (Dec. 1993), 3445–3462.

[Smi]   L. I. Smith, *A Tutorial on Principal Components Analysis*, http://csnet.otago.ac.nz/cosc453/student_tutorials/principal_ components.pdf.

[SN96]   Gilbert Strang and Truong Nguyen, *Wavelets and Filter Banks*, Wellesley-Cambridge Press, Wellesley, MA, 1996.

[Son06a]   Myung-Sin Song, *Wavelet Image Compression*, Ph.D. thesis, The University of Iowa, 2006.

[Son06b]   Myung-Sin Song, *Wavelet Image Compression*, Operator theory, operator algebras, and applications, Contemp. Math., vol. 414, Amer. Math. Soc., Providence, RI, 2006, pp. 41–73.

[Son07]   M. S. Song, *Entropy Encoding in Wavelet Image Compression*, Representations, Wavelets and Frames: A Celebration of the Mathematical Work of Lawrence Baggett (2007), 293–311.

[SR91]   Wim Sweldens and Dirk Roose, *Shape from Shading using Parallel Multigrid Relaxation*, Multigrid Methods, III (Bonn, 1990), Internat. Ser. Numer. Math., vol. 98, Birkhäuser, Basel, 1991, pp. 353–364. MR MR1131570 (92i:65190).

[Str92]   Jan-Olov Strömberg, *A Modified Franklin System as the First Orthonormal System of Wavelets*, Wavelets and Applications (Marseille, 1989), RMA Res. Notes Appl. Math., vol. 20, Masson, Paris, 1992, pp. 434–450. MR 1276533.

[Str97]   Gilbert Strang, *Wavelets from Filter Banks*, Springer, 1997.

[Str00]   Gilbert Strang, *Signal Processing for Everyone*, Lecture Notes in Math., vol. 1739, 2000.

[Swe96]   Wim Sweldens, The lifting scheme: A custom-design construction of biorthogonal wavelets, *Appl. Comput. Harmon. Anal.* **3** (1996), no. 2, 186–200. MR MR1385051 (97b:42060).

[Swe98]   Wim Sweldens, *The lifting scheme: A Construction of Second Generation Wavelets*, SIAM J. Math. Anal. **29** (1998), no. 2, 511–546 (electronic). MR MR1616507 (99e:42052).

[SWS90]   F. Schipp, W. R. Wade, and P. Simon, *An Introduction to Dyadic Harmonic Analysis*, Walsh series, Adam Hilger Ltd., Bristol, 1990, With the collaboration of J. Pál. MR 1117682 (92g:42001).

[Tha20]   Alaa Tharwat, *PCA (Principal Component Analysis) (https:// www.mathworks.com/matlabcentral/fileexchange/30792-pca-principal-component-analysis)*, MATLAB Central File Exchange (2020).

[Tre]   Brian F. Treadway, *Appendix to jor01*.

[Use01]   B. E. Usevitch, A tutorial on modern lossy wavelet image compression: Foundations of jpeg 2000, *IEEE Trans. Signal Processing* **18** (Sept. 2001), 22–35.

[Vet01]   M. Vetterli, Wavelets, approximation, and compression, *IEEE Signal Processing Magazine* **18** (Sept. 2001), 59–73.

[Wal99a]   J. S. Walker, *A Primer on Wavelets and their Scientific Applications*, Chapman Hall, CRC, 1999.

[Wal99b]   J. S Walker, *A Primer on Wavelets and their Scientific Applications*, Chapman & Hall, CRC, 1999.

[Wal02]   David F. Walnut, *An Introduction to Wavelet Analysis*, Applied and Numerical Harmonic Analysis, Birkhäuser Boston Inc., Boston, MA, 2002. MR MR1854350 (2002f:42039)

[Wan04]   M. S. Wang, Classical limit of von Neumann measurement, *Phys. Rev. A (3)* **69** (2004), no. 3, 034101, 2. MR MR2062693 (2005a:81037)

[Wat67]   Satosi Watanabe, *Karhunen-Loève Expansion and Factor Analysis: Theoretical Remarks and Applications*, Trans. Fourth Prague Conf. on Information Theory, Statistical Decision Functions, Random Processes (Prague, 1965), Academia, Prague, 1967, pp. 635–660. MR MR0234768 (38 #3084).

[WB03]   B. E. Wohlberg and C. M. Brislawn, Reversible integer-to-integer transforms and symmetric extension to even-length filter banks, *Visual Commun. & Image Process., Ser. Proc. SPIE* **5150** (July, 2003), no. 1, 1709–1718.

[Wel80]   R. O. Wells, Jr., *Differential Analysis on Complex Manifolds*, 2nd edition, Graduate Texts in Mathematics, vol. 65, Springer-Verlag, New York-Berlin, 1980. MR 608414.

[Wic94]   M. V. Wickerhauser, *Adapted Wavelet Analysis from Theory to Software*, IEEE Press, A K Peters, 1994.

Index